FLORA OF TROPICAL EAST AFRICA

BIGNONIACEAE

SALLY BIDGOOD, BERNARD VERDCOURT & KAJ VOLLESEN

Trees, shrubs or woody climbers, rarely (not native in Flora area) perennial or shrubby herbs. Leaves opposite, pinnate, bipinnate, digitately 2–3-foliolate or (not native in Flora area) simple, terminal leaflet sometimes replaced by a tendril; stipules absent but pseudostipules often developed and sometimes foliaceous. Flowers large and showy, bisexual, in several to many-flowered terminal or axillary racemes or panicles, or in fascicles, or solitary on dwarf shoots. Calyx gamosepalous, 5-toothed or -lobed to almost truncate or splitting spathaceously. Corolla gamopetalous, from almost symmetric to strongly bilaterally symmetrical; tube campanulate to infundibuliform or cylindrical; limb 5-lobed. Stamens 4, didynamous or (more rarely) subequal, adnate to the corolla tube, included or exserted, staminode often present; anthers 2-thecous (rarely 1-thecous), usually divergent, dehiscing longitudinally. Disk usually conspicuous, annular or tubular. Ovary superior, 2-carpellate, 2-locular with axile placentation or (*Kigelia*) unilocular with parietal placentation; ovules numerous in each locule; style one, terminal; stigma bilobed. Fruit a capsule, dehiscing by 2 loculicidal or septicidal valves, perpendicular or parallel to the septum, or (*Kigelia*) fleshy and indehiscent. Seeds numerous, compressed and winged in species with capsular fruits, wingless in species with indehiscent fruits.

About 110 genera and 900 species in all tropical regions but most numerous in Central and South America, with a few genera extending into subtropical and even (in eastern Asia) temperate regions. Represented by six native genera and ten native species in the Flora area.
Key to native and naturalised genera is given on page 7.

The family contains numerous species frequently cultivated as ornamentals. The following key, which was first prepared for the Flora Zambesiaca, has been modified to include further species grown in the Flora area. Species from the Flora Zambesiaca key not yet recorded for East Africa have been left in since they could well occur and since the key will then also be of use over a wider area. At least 37 species have been cultivated in East Africa apart from those naturalized.

KEY TO THE CULTIVATED SPECIES

1. Erect trees or shrubs (sometimes somewhat
   scandent in *Tecomaria*) 2
   Climbers, lianes or erect herbs 28
2. Leaves simple 3
   Leaves compound or sometimes simple
   and compound mixed 4
3. Leaves narrowly obovate to spathulate
   with attenuate base; corolla yellow with
   purple-brown markings; fruit spherical,
   up to 30 cm diameter . . . .   *Crescentia cujete* (p. 10)
   Leaves ovate-deltoid with truncate to
   subcordate base, corolla creamy white
   with yellow and purple markings, fruit up
   to 40 cm long, narrowly cylindrical,
   longitudinally ridged . . . .   *Catalpa bignonioides* (p. 9)

1

4. Simple and compound leaves together, the
   compound ones with 3(–5) leaflets and a
   broadly winged petiole; inflorescences
   cauliflorous on trunk and main
   branches, 1–2-flowered, flowers greenish
   with purple stripes; fruit ± globose,
   7–10 cm diameter . . . .   *Crescentia alata* (p. 10)
   Petioles not or narrowly winged; fruit not
   ± globose 5

5. Flowers solitary or 2–3 in fascicles on the
   old wood; corolla greenish white 6
   Flowers in terminal racemes or panicles 7

6. Branches usually spiny; fruits up to 17 cm
   long, curved, costate; corolla greenish
   white sometimes with purple markings . .   *Parmentiera aculeata* (p. 14)
   Branchlets not spiny; fruits 39–54 cm long;
   corolla white or greenish white . . . .   *Parmentiera cerifera* (p. 14)

7. Leaves pinnately 3-foliolate or palmately
   (1–)3–5(–9)-foliolate 8
   Leaves 1–3-pinnate 15

8. Leaves pinnately 3-foliolate; flowers yellow   forms of *Tecoma stans* (p. 26)
   Leaves palmately (1–)3–5(–9)-foliolate
   (*Tabebuia*) 9

9. Leaflets pubescent or hairy at least on veins
   or in nerve-axils beneath 10
   Leaflets glabrous beneath (but may be
   lepidote) 13

10. Corolla pinkish purple to deep magenta,
    pubescent outside; fruit glabrous . . . . .   *Tabebuia impetiginosa* (p. 18)
    Corolla yellow, glabrous or with sparse
    hairs on veins 11

11. Leaflets hairy only in nerve axils beneath . .   *Tabebuia guayacan* (p. 17)
    Leaflets more extensively hairy beneath 12

12. Calyx and fruit with dense long hairs . . .   *Tabebuia chrysotricha* (p. 17)
    Calyx tomentose with shorter indumentum;
    fruit glabrous or glabrescent . . . .   *Tabebuia chrysantha* (p. 17)

13. Flowers bright yellow . . . .   *Tabebuia aurea* (p. 17)
    Flowers white to pinkish purple or magenta,
    sometimes with a yellow throat 14

14. Leaflets elliptic with long-acuminate to
    cuspidate apex and obtuse to acute
    base; corolla white to pinkish purple or
    magenta . . . .   *Tabebuia rosea* (p. 18)
    Leaflets often more obovate but very
    variable, obtuse to rounded at the apex
    and base; corolla lavender, pale magenta
    or mauve with yellow throat . . . .   *Tabebuia heterophylla* (p. 18)

15. Leaves once pinnate 16
    Leaves bipinnate or tripinnate 23

16. Calyx spathaceous 17
    Calyx campanulate or cylindrical 21

17. Calyx pubescent to densely tomentose 18
    Calyx glabrous or glabrescent 20

18. Flowering calyx usually lepidote with a few
    scattered hairs, not densely covered with
    multicellular hairs; corolla yellow; fruit
    linear 35–105 cm long, frequently falcate,
    pendulous .... *Markhamia lutea* (p. 34)
    Flowering calyx densely covered with
    multicellular hairs 19
19. Corolla yellow; fruit linear, (20–)39–87 cm
    long, pendulous .... *Markhamia obtusifolia* (p. 35)
    Corolla orange to red; fruit fusiform,
    15–23 cm long, held erect .... *Spathodea campanulata* (p. 29)
20. Corolla campanulate, 5-lobed, yellow,
    frequently with maroon or brown spots;
    calyx glabrescent, the surface usually
    lepidote; fruit narrow, compressed,
    14–72 cm long .... *Markhamia zanzibarica* (p. 32)
    Corolla unequally 5-lobed, white, the tube
    long, narrowly cylindrical; calyx glabrous;
    fruit compressed, 17–38(–50) cm long .. *Dolichandrone alba* (p. 10)
21. Calyx campanulate, 1.4–2.1 × 1.7–2.2 cm,
    the lobes rounded to obtuse; leaflets
    entire (rarely crenate); corolla broadly
    campanulate, red, frequently with yellow
    in the throat; fruits narrow, up to 54 cm
    long, compressed, loosely spiralled .... *Fernandoa magnifica* (p. 42)
    Calyx narrowly campanulate to cylindrical,
    0.3–1.7 × 0.2–0.8 cm; leaflets crenate or
    serrate 22
22. Leaflets crenate; corolla orange or red,
    rarely yellow, strongly bilabiate with
    anthers exserted, curved, 3.5–6(–6.5)
    cm long of which the tube is 2–3.5 cm;
    anther thecae connate for upper third;
    fruit narrow, compressed, attenuate at
    both ends, 4–13(–19.5) cm long – 2
    subspecies. See key on p. 24 .... *Tecomaria capensis* (p. 24)
    Leaflets serrate; corolla yellow, ± regular,
    not strongly bilabiate; anthers ±
    included, straight, 3.5–5.5 cm long,
    narrowly campanulate with the tube
    constricted above the calyx for 0.3–0.5
    cm, or infundibuliform; anther thecae
    not connate; fruit narrow, compressed,
    7–21 cm long. 3 species. See key on p. 26   *Tecoma* (p. 26)
23. Leaves large, 0.9–1.8 m long with large
    leaflets; corolla reddish purple to
    maroon, tube 5–7 cm long, yellow to pink
    inside; fruit sword-like, 50–100 × 7–9 cm   *Oroxylum indicum* (p. 12)
    Leaves much smaller (except in *Millingtonia*
    which has white flowers); fruits much
    narrower, or if not, then round 24
24. Corolla blue-purple; fruits compressed
    discoid or broadly oblong-elliptic; leaflets
    numerous, (9–)13–41 per pinna 25
    Corolla white or pale yellow green; fruits
    elongate; leaflets fewer 26

25. Leaves with 13–31 pinnae; basal part of corolla tube almost straight, upper part funnel-shaped; fruit discoid, with strongly wavy edges .... *Jacaranda mimosifolia* (p. 11)

Leaves with 8–12 pinnae; basal part of corolla tube sharply bent and constricted above base, upper part campanulate; fruit oblong-elliptic with much straighter edges .... *Jacaranda obtusifolia* (p. 11)

26. Domatia present on the underside of the leaflets; calyx less than 0.4 cm long; corolla white, tube very long and narrow, widening abruptly at the lobes; fruit narrow, compressed, 22–35 × 1.7–2.8 cm *Millingtonia hortensis* (p. 12)

Domatia absent on the underside of the leaflets; calyx 0.7–3.5 cm long; corolla yellow, white or very pale yellow-green within; fruit cylindrical, up to 68 cm long 27

27. Calyx 1.8–4 cm long; corolla (7.2–)8–13.5 (–15) cm long, tube infundibuliform; fruit smooth .... *Radermachera sinica* (p. 16)

Calyx 0.7–1.5 cm long; corolla 3.5–5.5 (–7) cm long, tube campanulate; fruit verrucose .... *Radermachera xylocarpa* (p. 16)

28. Erect herb 20–30(–90) cm tall with pinnate leaves having 6–11 pairs of leaflets; inflorescence 2–10-flowered; corolla pinkisk purple and yellow outside, yellow with purple lines inside, tube 4.5–6 cm long; fruit quadrangular, 5–7.5 cm long, slightly curved .... *Incarvillea delavayi* (p. 11)

Climbers or lianas 29

29. Slender climber with bipinnate leaves terminating in a branched tendril; corolla orange to scarlet, narrowly tubular, ± 2.5 cm long with very short limb; fruit ellipsoid, ± 4 cm long, smooth *Eccremocarpus scaber* (p. 10)

Not as above 30

30. Slender annual creeper; leaves with 2 pinnae, each pedately 5-foliolate with branched tendril between; flowers in subspicate inflorescences; corolla purple or green with tube 1 cm long and lobes fused to form a lip 6–8 mm long; fruit ovoid, 3–3.5 cm long, densely covered with hooked bristles .... *Tourettia lappacea* (p. 21)

Without any of the above characters 31

31. Leaflets (3–)5–15, tendrils absent 32

Leaflets 1–3 (rarely also with 2 stipule-like leaflets at the base of the petiole), tendrils often present replacing terminal leaflet 37

32. Corolla 5–11.5 cm long, climbers with aerial roots 33

Corolla 1–6 cm long 35

33. Leaflets entire; corolla infundibuliform, pink, often cream on the lobes, (7–)8–11 cm long; inflorescences usually cauliflorous on old wood, sometimes on young foliage shoots; calyx (1–)1.5–2.5 cm long, the lobes 0.4–0.6 cm long; leaflets 3–13 × 1.5–7 cm; fruit woody *Tecomanthe dendrophila* (p. 21)
    Leaflets serrate or dentate; corolla orange-red to red, 5–8 cm long; inflorescence terminal 34
34. Inflorescence a raceme; calyx 1.7–2.5 cm long, lobes under 1 cm long, acute; corolla tube 5–7 cm long, lobes 1–1.5 × 1.5–2 cm . . . . *Campsis radicans* (p. 9)
    Inflorescence a narrow panicle; calyx 2.2–3.2 cm long, lobes over 1 cm long, cuspidate; corolla tube 3.5–5 cm long, lobes 1.5–2.5 × 2.5–3.5 cm . . . . *Campsis grandiflora* (p. 9)
35. Calyx 0.7–2.2 cm long with lobes up to 0.9 cm long; corolla broadly campanulate, 3.5–6.5 cm long, pink, yellowish with dark pink guide lines in the throat; fruit 30–45 cm long, cylindrical, compressed, apiculate . . . . *Podranea ricasoliana* (p. 14)
    Calyx 0.2–1 cm long, not obviously lobed; corolla 1–6 cm long; fruit 5–12 cm long, ovoid-ellipsoid, acuminate 36
36. Calyx less than 0.4 cm long, corolla 1–3.5 cm long, creamy yellow with purple to brown markings on the inside *Pandorea pandorana* (p. 12)
    Calyx 0.7–1 cm long; corolla 5–6 cm long, white, pink or mauve, frequently with dark pink to crimson markings into throat . . . . *Pandorea jasminoides* (p. 12)
37. Calyx densely tomentose, the surface completely covered with short velvety hairs 38
    Calyx glabrous to pubescent or pilose, the surface never completely covered in hairs 40
38. Corolla white, yellow or cream, 3–6.5 cm long 39
    Corolla mauve to red or yellow with red markings and purplish lobes, yellow in throat, 9–10 cm long; fruit ellipsoid, obtuse at either end, 14–18.5 × 4.8–7 cm *Distictis buccinatoria* (p. 10)
39. Calyx with large glands; corolla yellow, ± 5 cm long; fruit 18–28 cm long, up to 2.8 cm wide, not echinate . . . . *Adenocalymma marginatum* (p. 8)
    Calyx without large glands; corolla cream outside, yellow inside, 3–5 cm long; fruits up to 24(–31) × 5–7(–7.5) cm, echinate *Pithecoctenium crucigerum* (p. 14)
40. Inflorescences terminal 41
    Inflorescences axillary 45

41. Corolla orange, yellow or red, 4–8 cm long,
    tubular, densely tomentose on margins;
    lobes narrowly oblong to oblong; stamens
    exserted; calyx 0.4–0.7 cm long . . . .   *Pyrostegia venusta* (p. 16)
    Corolla pink, purple or mauve 42
42. Leaflets obovate, cuneate at the base,
    obtuse at the apex; calyx 0.7–1.2 cm long,
    glabrous; corolla 5.5–9 cm long, tubular-
    campanulate, the lobes obovate, pink,
    mauve, magenta or purple, throat white
    or yellow, with or without purple lines . .   *Saritaea magnifica* (p. 16)
    Leaflets ovate, cordiform or elliptic (obovate
    in *Clytostoma calystegioides*), rounded,
    truncate or subcordate at the base,
    rounded or acuminate at the apex 43
43. Leaflets ovate or cordiform; calyx 0.7–1.2 cm
    long; corolla 2–4.5 cm long, campanu-
    late, pinkish mauve, glandular-pubescent
    on the outside, lobes obovate; stamens
    included . . . .   *Arrabidaea selloi* (p. 8)
    Leaflets elliptic 44
44. Calyx with subulate teeth; flowers usually
    paired or few per inflorescence; corolla
    mauve to purple, lighter at throat . . . . .   *Clytostoma calystegioides* (p. 9)
    Calyx truncate or with minute denticles;
    flowers 1–4, usually axillary but can
    appear terminal; corolla lavender, pink
    or pale purplish with white throat . . . . .   *Clytostoma binatum* (p. 9)
45. Corolla pink to purple or mauve 46
    Corolla yellow or orange to orange-red 47
46. Pods with linear valves, ridged and with a
    raised central rib . . . .   *Mansoa difficilis* (p. 11)
    Pods narrowly ellipsoid, echinate . . . .   *Clytostoma binatum* (p. 9)
47. Pods with linear valves, ridged and with a
    raised central rib . . . .   *Mansoa difficilis* (p. 11)
    Pods not as above 48
48. Tendrils claw-like; calyx broadly campanu-
    late, 1–1.2 × 1.5–1.7 cm; leaflets small,
    (2.7–)3.1–5.3(–5.6) × (0.8–)0.9–1.8
    (–2.3) cm; corolla yellow, 5.8–10 cm
    long and 5–7 cm across the lobes; lobes
    very broadly obovate, ± half the length
    of the tube; fruit narrow, 39–153 cm ×
    0.5–1.5 cm, drying dark brown . . . .   *Macfadyena unguis-cati* (p. 22)
    Tendrils not claw-like; calyx 0.8–1(–1.2) ×
    0.8 cm 49
49. Stamens exserted; corolla lobes narrowly
    oblong to oblong, densely tomentose on
    the margins; fruit up to 33 cm long,
    narrowly compressed . . . .   *Pyrostegia venusta* (p. 16)
    Stamens included; corolla lobes broadly
    oblong or very broadly so, glabrous or
    with scattered hairs on the surface and
    margins 50

50. Leaflets subcordate to cordate at the base;
fruit ± 19 cm long, compressed; flowers
on short axillary shoots in leaf axils ...   *Bignonia capreolata* (p. 9)
Leaflets cuneate, truncate or rounded at
the base; fruit ellipsoid, 6–12 cm long;
flowers in axillary clusters or long
racemes 51
51. Flowers in elongated 4–8-flowered axillary
racemes 8–20 cm long ....   *Anemopaegma chamberlaynii* (p. 8)
Flowers 2–4, in very short clusters 52
52. Stems and leaves densely puberulous;
venation of leaflets characteristically
impressed above on dry leaves ....   *Anemopaegma rugosum* (p. 8)
Stems and leaves glabrous; venation not
impressed on dry leaves ....   *Anemopaegma chrysoleucum* (p. 8)

KEY TO NATIVE AND NATURALISED GENERA

1. Leaves bifoliolate with branched claw-like tendrils;
woody climber ....   1. **Macfadyena** (p. 22)
Leaves pinnate without tendrils; trees and shrubs 2
2. Flowers in pendulous lax panicles up to 1(–1.5) m
long; fruit fleshy, sausage-shaped, up to 1 m long,
indehiscent; seeds not winged; ovary 1-locular ..   8. **Kigelia** (p. 43)
Flowers in erect shorter panicles or racemes,
fascicles or sometimes solitary; fruit a capsule;
seeds winged, ovary 2-locular 3
3. Calyx spathaceous 4
Calyx not spathaceous, lobed or toothed 6
4. Corolla pink, mauve or purple; calyx glabrous ..   **Newbouldia** (p. 21)
Corolla yellow to orange, red or reddish purple 5
5. Calyx densely covered with multicellular hairs,
corolla orange to red (rarely yellow), fruit
fusiform, 15–29 cm long, held erect ....   4. **Spathodea** (p. 29)
Calyx scaly or with few scattered hairs (tomentose
in *M. obtusifolia*); corolla yellow, often with
brown spots, or reddish purple; fruit narrow and
compressed, 14–80(–105) cm long, pendent ...   5. **Markhamia** (p. 32)
6. Corolla pink to purple, 2.3–6.5 cm long; flowers
in large terminal panicles ....   6. **Stereospermum** (p. 37)
Corolla pure yellow or orange to crimson with a
yellow throat 7
7. Corolla curved, narrowly funnel-shaped, orange
to scarlet (rarely yellow); stamens well exserted;
anther thecae connate for the upper third ...   2. **Tecomaria** (p. 24)
Corolla not curved, much broader (save in one
cultivated *Tecoma*); stamens included or if slightly
exserted then thecae not connate 8
8. Flowers in axillary clusters or racemes or in
clusters on the old wood or solitary; corolla
broadly campanulate, orange to crimson with
yellow throat or pure yellow; native trees .....   7. **Fernandoa** (p. 40)
Flowers in terminal racemes or panicles; corolla
tubular to campanulate, usually bright yellow;
cultivated or naturalised small trees or shrubs .   3. **Tecoma** (p. 26)

CULTIVATED SPECIES

Species belonging to genera which contain native or naturalised species will be found in the accounts of those genera. Jex-Blake, Gard. in E. Afr. ed. 4 (1957)* refers to about 30 cultivated species but no material has been seen of some of these. They have, however, all been included in the key and briefly dealt with.

Adenocalymma marginatum (*Cham.*) *DC.* (*A. nitidum* sensu Jex-Blake: 131). Note: the genus name has been spelled as both *Adenocalymna* and *Adenocalymma*. The latter spelling has been conserved over the former.

Fig. 3.5–3.8, p. 19

Extensive climber. Leaves with one pair of leaflets; tendril often present in position of terminal leaflet; leaflets coriaceous, elliptic, ± 11 × 7 cm, with reflexed margins. Calyx ± 7 mm long, with large dark glands; corolla golden yellow, narrowly funnel-shaped, tube ± 5 cm long; lobes 1.7 × 1.4 cm. Fruit 18–28 × 2.8 cm.

Native of S America. Cultivated in Kenya: Nairobi, Nairobi Arboretum XXXI, 10 Apr. 1952, *Dyson* 305; Kilifi District: Malindi, Casuarina Point, in garden, 10 Jan. 1988, *Robertson* 5090.

Anemopaegma chamberlaynii (*Sims*) *Bur. & K.Schum.* (*A. racemosum* Mart., *Bignonia chamberlaynii* Sims; Jex-Blake: 132)

Rather delicate climber with long thin unbranched stems to 6 m tall. Leaves with one pair of leaflets, sometimes with pseudostipules; leaflets ovate to lanceolate, 6.5–9 × 2.5–5.5 cm, acute at the apex. Racemes axillary, elongated, 8–20 cm long, 4–8-flowered, sometimes with leafy branches at lower nodes; calyx 6 × 8 mm, subtruncate; corolla creamy or primrose yellow, campanulate to funnel-shaped, narrowed at base into short to long tube, 6–7 cm long, the lobes 1 × 1.3 cm. Fruit woody, compressed ellipsoid, ± 9 × 6 cm; seeds ± elliptic, including the very thin wing ± 5 × 4 cm.

Native of Brazil. No East African material has been seen.

Anemopaegma chrysoleucum (*Kunth.*) *Sandw.* (Jex-Blake: 132 as *Bignonia chrysoleuca* Kunth.)

Rampant climber or slender liane rooting wherever it touches the ground. Leaves with one pair of leaflets, with or without a terminal tendril; leaflets elliptic, oblong-elliptic or lanceolate, 5–25 × 2–16 cm, acute or acuminate; veins prominent or rarely impressed; pseudostipules usually evident, round, 0.5–2 cm diameter. Inflorescences axillary, 1–3 or more-flowered but can appear ± terminal; calyx 7–12 mm long with broad shallow lobes; corolla with pale yellow tube and ± white lobes; tube 4–7(–9) cm long, narrowed at base for 1.2–3 cm and 1.5–3 cm wide at the throat; lobes rounded, ± 10 × 12 mm. Fruit ellipsoid, 6–12 × 3–6 cm; seeds not winged.

Native from Mexico to Brazil. No East African material has been seen.
NOTE. Jex-Blake: 132 mentions a *Bignonia albolutea* being very similar but we have not traced this.

Anemopaegma rugosum *Schltdl.*

Densely hairy climber. Leaves with one pair of leaflets and leaf-like pseudo-stipules; leaflets ovate, ± 11 × 7 cm, shortly acuminate at the apex, rounded at the base, the venation above characteristically impressed on dry specimens and surface rugose; tendrils 3-fid (fide Gillett) arising from apex of petiole. Flowers in axillary clusters, the pedicels ± 7 mm long; calyx ± 8 mm long, truncate; corolla primrose yellow, narrowly funnel-shaped, tubular at base, ± 5 cm long. Fruit 10–12 cm long.

Native of Venezuela. Cultivated in Kenya: Nairobi, Langata, Hort. P. Greensmith, 2 Dec. 1975, *Gillett* 20916 – a sterile specimen but it matches very well with the original material used to prepare Bot. Mag. 116: t. 7124 (1890).

---

* Since there are numerous references to this it has been abbreviated to Jex-Blake in this list.

Arrabidaea selloi (*Spreng.*) *Sandw.*
Woody climber. Leaves with single pair of leaflets and with or without a terminal leaflet, no tendrils; leaflets ovate or cordiform, to 10 × 7 cm (terminal to 13 × 9). Flowers in a terminal panicle; calyx 0.7–1.2 cm long; corolla 2–4.5 cm long, campanulate, pinkish mauve, glandular pubescent on the outside, lobes obovate; stamens included. Fruit linear, straight, 25–35 × 1–2 cm.
Native of South America. No material has been seen from East Africa, but the species has been cultivated in Zimbabwe.

Bignonia capreolata *L.* Cross Vine, Trumpet Flower.
Glabrous or subglabrous climber. Leaves with one pair of leaflets and cordiform pseudo-stipules; leaflets narrowly ovate-cordiform, ± 10 × 4 cm, acute with obtuse tip, subcordate to cordate at the base; tendrils 3-fid, from apex of petiole. Flowers in clusters of 1–3 on short-shoots in leaf axils; calyx ± 7 mm long, with broadly triangular teeth; corolla orange-red with reddish-yellow lobes, tube ± 4 cm long, tubular for basal 1 cm, funnel-shaped above. Fruit ± 19 cm long (only one seen), linear, compressed.
Native of North America. Cultivated in Tanzania: Lushoto District: Amani Nursery, 28 Sept. 1943, *Greenway* 6818.

Campsis grandiflora (*Thunb.*) *K.Schum.* (Jex-Blake: 141 as *Tecoma grandiflora* Thunb.)
Glabrous or subglabrous climber. Leaves pinnate; leaflets 5–9, ovate to elliptic, to 8 × 5 cm, acute to cuspidate, coarsely dentate. Inflorescence a narrow panicle; calyx 2.2–3.2 cm long, lobes over 1 cm long, cuspidate; corolla orange red to red, tube 3.5–5 cm long, lobes 1.5–2.5 × 2.5–3.5 cm. Fruit not seen.
Native of China. No East African material has been seen.

Campsis radicans (*L.*) *Seem.* Trumpet Creeper. (Jex-Blake: 141 as *Tecoma radicans* L.). Jex-Blake mentions hybrids between the two species
Glabrous or subglabrous climber. Leaves pinnate; leaflets 7–13, ovate to elliptic, to 8 × 4.5 cm., acute to cuspidate, almost entire to coarsely dentate. Inflorescence a terminal raceme; calyx 1.7–2.5 cm long, lobes under 1 cm long, acute; corolla orange-red to red, tube 5–7 cm long, lobes 1–1.5 × 1.5–2 cm. Fruit curved, narrowly ellipsoid, not flattened, with thick woody valves, 12–15 × 2–3 cm; seed ellipsoid, ± 2 × 0.5 cm including wings.
Native of North America. No East African material has been seen.

Catalpa bignonioides *Walt.* (Jex-Blake: 107). Indian Bean Tree.
Tree to 10 m tall with rounded crown. Leaves simple, large, ovate; leaflets 7–30 × 7–30 cm, with subcordate or truncate base. Corolla white with yellow and purple markings. Fruit narrowly cylindrical, up to 40 cm long, ridged.
Native of North America. No material has been seen from East Africa.

Clytostoma binatum (*Thunb.*) *Sandw.* (*C. purpureum* (Hook. f.) Rehd.; Jex-Blake: 132, 354 as *Bignonia purpurea* Hook. f.)
Climber with long slender stems. Leaves with 2 leaflets and sometimes tendrils; leaflets elliptic to oblong-ovate, 6–15 × 3.5–6.5 cm, shortly to long-acuminate, entire. Flowers 1–4, axillary or appearing terminal; calyx 4–6 mm long, truncate or minutely denticulate; corolla lavender, pink or pale purplish with white centre; tube 2.5–5.5 cm long; lobes ovate, 1–2.5 × 1.5–2 cm. Fruit ellipsoid, 4–7 × 1.8–2.8 cm, densely echinate; seeds discoid-cordiform, 2.5–3.5 × 1.2–3 cm.
Native of Central and Northern America. No material has been seen from East Africa.

Clytostoma calystegioides (*Cham.*) *Baill.* (Jex-Blake: 132 as *Bignonia speciosa*)
Climbing shrub. Leaves with paired leaflets and sometimes a terminal purplish tendril; leaflets oblanceolate to obovate or elliptic, 3–9.5 × 1.2–4.3 cm. Flowers paired or few; calyx ± 6 mm long, with distinct filiform teeth; corolla pale lavender-purple with darker veining, yellowish inside tube, narrowly funnel-shaped, 6–7 cm long; lobes rounded and undulate. Fruit oblong, 7.5 × 3 cm, echinate.

Native of South America. Cultivated in Kenya: Cherangani Hills, Hort. A. Barnby, 24 Feb. 1947, *Jex-Blake* H20/47; Nairobi, Chiromo Campus, 1 Dec. 1971, *Mathenge* 782; and in Tanzania: Amani Nursery, 29 Oct. 1948, *Greenway* 8319.

Crescentia alata *Kunth.* Calabash Tree (U.O.P.Z.: 214 (1949) pro parte as *C. cujete*)
Small tree 3–8 m tall. Leaves simple and compound, the latter 3(–5)-foliolate; petioles winged; leaflets oblanceolate, 1–4.5 × 0.3–1.2 cm. Inflorescences borne on trunk, 1–2 flowered; calyx split into two lobes 1.4–1.9 × 0.7–1.4 cm; corolla dark red, campanulate, fleshy, constricted below the limb, 4–6 cm long. Fruit spherical or ovoid, 7–10 cm in diameter.
Native from Mexico to Costa Rica. Cultivated in Tanzania: Dar es Salaam Botanic Gardens, 21 Nov. 1972, *Ruffo* 575.

Crescentia cujete *L.* Calabash tree (T.T.C.L.: 70 (1949); U.O.P.Z.: 214 (1949) pro parte; Dale, List Introd. Trees Uganda: 25 (1953))
Tree to 10 m tall. Leaves tufted, simple, sessile, obovate-spathulate, 4–26 cm long. Inflorescences 1–2-flowered, borne on trunk or main branches; calyx split into two lobes 1.8–2.6 × 1.3–2.4 cm; corolla yellow-green, speckled red or marked purple, campanulate, fleshy, 4–7.5 cm long with transverse fold across the tube. Fruit spherical to ovoid or ellipsoid, 15–30 cm long, (8–)13–20 cm diameter with thin hard shell.
Native to Mexico and North Central America but natural range obscured by very extensive cultivation in tropical America. Cultivated in Kenya: Kwale District: Diani Beach, Kivulini, 25 Feb. 1992, *Luke* 3075 and Tanzania: Pangani District: Mwera, 15 July 1932, *Geilinger* 865; Zanzibar, Dunga, Feb. 1929, *Greenway* H 11/29.

Distictis buccinatorius (*DC.*) *Gentry* (Jex-Blake: 132, as *Bignonia buccinatoria* DC.; *Phaedranthus buccinatorius* (DC.) Miers; *Bignonia cherere* Lindl.)   Fig. 4.1–4.4, p. 20.
Climber with 4-angular stems. Leaves with 2 leaflets and tendrils arising between each pair; leaflets ovate, oblong or elliptic, up to 12 × 6.5 cm. Flowers in long narrow panicles at ends of branchlets; calyx ± 1 cm long; corolla rich wine-coloured with crimson to purple lobes and tube orange-yellow or golden yellow at base, narrowly funnel-shaped with tubular base, 9–10 cm long, with rather short broad lobes. Fruit ellipsoid, 14–18.5 × 4.8–7 cm.
Native of Central America. Cultivated in Kenya: Nairobi Arboretum, 10 Apr. 1952, *Dyson* 341; Nairobi, Loresho ridge, 8 Oct. 1977, *Gillett & Stearn* 21586; Kiambu District: Muguga, Hort. Greenway, 12 Oct. 1963, *Greenway* 10900) and Tanzania: Amani, Bustani, 28 Sept. 1943, *Greenway* 6817; Amani Nursery, 6 Dec. 1945, *Fernie* s.n.

Dolichandrone alba (*Sim*) *Sprague*
Shrub or small tree to 12 m tall. Leaves 1-pinnate, 3–4(–5)-jugate; leaflets ovate or elliptic, 1.5–7.5(–8.5) × 1–4(–5) cm. Flowers in terminal racemes or panicles to 50 cm long; calyx spathaceous, 2–3.3 cm long, glabrous; corolla white, unequally 5-lobed, tube 3.5–5.5 cm long, narrowly cylindrical, lobes 2–3 × 1.7–2.5 cm. Fruit linear, compressed, curved, 17–38(–50) × 1.3–2.8 cm.
Native of Mozambique. No material has been seen from East Africa, but the species has been cultivated in Mozambique and South Africa.

Eccremocarpus scaber *Ruiz & Pav.* Chilean Glory (Jex-Blake: 135)
Slender climber to 3 m. Leaves bipinnate, terminating in a branched tendril; leaflets rounded or obliquely cordate, 0.8–2 cm long, entire or serrate, obtuse. Flowers in terminal racemes 10–15 cm long; corolla orange to scarlet, narrowly tubular, ± 2.5 cm long with very short limb. Fruit ellipsoid or elongate ovoid, ± 4 cm long; seeds with a circular wing.
Native of Chile. Apparently rarely grown in Kenya but no material has been seen.

Fernandoa. See page 40.

Incarvillea delavayi *Bur. & Franch.* (Jex-Blake: 87)

Herb, 20–30(–90) cm tall. Leaves pinnate, leaflets in 6–11 pairs, lanceolate, 1.3–5 × 0.5–2.5 cm, crenate. Racemes 2–10-flowered; corolla with tube purple and yellow outside, yellow lined with purple inside and lobes purple or magenta, tube narrowly funnel-shaped, ± 6 cm long; lobes ± 3 × 2.5 cm. Fruit ± 7.5 × 1.5 cm; seeds winged.

Native of Eastern Himalayas. No East African material has been seen.

Jacaranda mimosifolia *D.Don* (Jex-Blake: 116, 355 as *J. mimosaefolia,* Jex-Blake: 331 as *J. ovalifolia* R.Br.; T.T.C.L.: 70 (1949); Dale, List. Introd. Trees Uganda: 46 (1953); Liben in F.A.C. Bignoniaceae: 34 (1977); Tardelli & Settesoldi in Fl. Som. 3: 307, fig. 212 (2006)   Fig. 6.8–6.11, p. 27.

Tree up to 15 m or occasionally a shrub 2.5–3 m tall. Leaves bipinnate with 13–31 pinnae each with 13–41 sessile leaflets which are elliptic or oblong-elliptic, 3–12 × 1–4 mm, but terminal leaflet can be up to 25 × 7 mm; rachis unwinged. Flowers in terminal panicles; calyx campanulate, 3–4 mm long with teeth ± 1 mm long; corolla purplish blue or lilac, the tube white inside, basal part almost straight, upper part funnel shaped, 3–4 cm long with lobes 3–8 mm long. Fruit round, 3.2–6 × 3.7–6 cm, strongly compressed, woody; seeds including the surrounding wing 9–17 × 11–17 mm.

Native of Argentina and Bolivia. Widely cultivated in East Africa in gardens and as a street tree. Uganda: Mbale District, Mt. Elgon Hotel, 25 Jan. 1971, *Wendelberger* U166. Kenya: Nairobi, Karura Forest, 25 Nov. 1966, *Perdue & Kibuwa* 8096. Tanzania: Arusha, 20 Oct. 1965, *Leippert* 6122; Lushoto District: Amani, Rest House Garden, 1 Nov. 1928, *Greenway* 965; Tabora, near Moravian Church, 12 Nov. 1977, *Ruffo* 946.

NOTE. Williams in U.O.P.Z.: 313 mentions *J. caerulea* (L.) Griseb. Stating 'there are a number of young trees in Zanzibar town which have flowered but sparingly'. There is nothing in the information given to suggest this is a correct identification. A native of the West Indies, *J. caerulea* differs widely from *J. mimosifolia* in its oblong-elliptic, ± obtuse leaflets (0.5–)1–2 × (0.3–)0.5–1 cm and more elliptic fruit. It is, however possible that the original determination is correct since it is likely a West Indian species would be thought to be more suited to Zanzibar climate than *J. mimosifolia* but there is no record of it being cultivated anywhere outside America.

Jacaranda obtusifolia Humb. & Bonpl. (*J. lasiogyne* Bur. & K.Schum.)

Shrub or tree to 15 m tall. Leaves bipinnate with numerous leaflets, pinnae 8–12, leaflets obtuse, 2–17 × 1–7 mm, terminal up to 35 × 12 mm; rachis unwinged. Flowers in terminal panicles; calyx broadly campanulate, 2–3 mm long with teeth ± 0.5 mm long; corolla purple, the tube white inside, basal part sharply bent and constricted above base, upper part campanulate, 3–4 cm long with lobes up to 1 cm long. Fruit oblong-elliptic, parallel-sided, 4–6.6 × 2–3.5 cm.

Native from Venezuela to Bolivia. Has been cultivated in Tanzania: Songea District, Mahenge, 10 Jan. 1931, *Schlieben* 1615.

Macfadyena. Since this appears to be, at least occasionally, naturalised it is treated fully; see page 22.

Mansoa difficilis (*Cham.*) *Bur. & K.Schum.* (*Chodanthus praepensus* (Miers) Summerhayes)

Climber. Leaves with one pair of leaflets and terminal 3-fid tendrils; leaflets broadly lanceolate to elliptic, 3.5–8 × 1.3–4.5 cm. Flowers 1–3 in leaf-axils; calyx ± 5 mm long; corolla orange-red outside, flushed orange-red inside and with reddish yellow lobes or pink to purple, funnel-shaped, ± 5 cm long. Fruit linear-oblong, ± 15.5 × 1.5–2 cm, longitudinally ridged.

Native of Brazil. Cultivated in Kenya: Nairobi, Mar. 1953, *Jex-Blake* H82/53/4; Nairobi Arboretum Block XXXII, 15 Apr. 1952, *Dyson* 261 & 16 June 1952, *G. R. Williams* 447; and Tanzania: Lushoto District: Amani Nursery, 28 Sept. 1943, *Greenway* 6818.

Markhamia. Cultivated material is mentioned under the native species, see page 32.

Millingtonia hortensis *L.f.* Indian Cork Tree (U.O.P.Z.: 352, fig. (1949); I.T.U.: 51 (1953))   Fig. 1.1–1.4, p. 13.

Tree up to 18(–25) m tall. Leaves large, bipinnate, up to 1 m long, the pinnae in 2–3 pairs each with 1–2 pairs of leaflets and a terminal leaflet, leaflets ovate to lanceolate-elliptic, 6–7 × 3–5 cm. Flowers in terminal panicles 10–40 cm long, fragrant, opening mostly at night; calyx campanulate, ± 4 mm long; corolla white, tube 6–8 cm long, lobes ovate, 15–20 × 8 mm. Fruit linear, 22–35 × 1.7–2.8 cm, flattened.

Native of Southeast Asia and Indonesia. Cultivated in Uganda: Entebbe Botanic Gardens, fide Dale; Kenya: Kwale District: Diani Beach, 7 July 1971, *Dyson* 629; Kilifi District: Kibarani, 30 May 1945, *Jeffery* 213 & Malindi, Dec. 1962, *Dale* K 2024; and Tanzania: Tanga District: Pangani, by river, 1 Mar. 1961, *Verdcourt* 3099 & Tanga, 29 May 1961, *Denning* s.n.; Morogoro, May 1947, *Wigg* FH2232; a common street tree in Dar es Salaam.

Oroxylum indicum *(L.) Vent.* is mentioned in U.O.P.Z.: 393 (1949).

Small tree 5–8(–13) m tall with thick bark and large corky lenticels. Leaves large, 0.9–1.8 m long, 2–3-pinnate; leaflets ovate-elliptic, (7–)9–16 × 3–9 cm, long, obtuse to cordate at the base, acuminate at the apex. Inflorescences on thick branch-like peduncles, persistent, ± 35 cm long; pedicels 3–7 cm long; calyx blackish purple, 3–4.5 cm long; corolla deep maroon to reddish purple outside, yellowish or pinkish inside; tube 5–7 cm long, 3 cm wide at throat; lobes obovate, 4–6 × 3–4.5 cm. Fruit sword-like, 50–75(–100) × 7–9 cm, flat, narrowed to both ends; valves ± woody with mid- and marginal ridges; seeds elliptic, ± 6.5 × 3 cm including the very thin membranous wing.

Native of India, Ceylon, Southeast Asia, Indonesia and the Philippines. Known from a single specimen only in the Migombani Gardens, Zanzibar. We have seen no material and do not know if this tree is still extant.

Pandorea jasminoides *(Lindl.) K.Schum.* Bower of Beauty (Jex-Blake: 141, as *Tecoma jasminoides,* he also mentions var. *alba* and var. *rosea*)

A vigorous climber to 5 m. Leaves very glossy, pinnate, ± elliptic, 2–5 × 1–3 cm, coriaceous. Flowers few in terminal panicles; calyx 7–10 mm long with short obtuse lobes; corolla white with crimson or purple throat or in some varieties pure white or rose; tube 3–4(–7.5) cm long, 5 cm wide at throat with round lobes 2 × 2 cm. Fruit narrowly oblong.

Native of Northeast Australia. Cultivated in Kenya: Trans-Nzoia District: Kitale, Kaporetwa, Hort. T.H.E. Jackson, 30 Sept. 1959, *Verdcourt* 2446; Nairobi, Mar 1953, *Jex-Blake* H82/53/2 ("var. *alba*").

Pandorea pandorana *(Andr.) Steenis* (Jex-Blake: 141 as *Tecoma australis* R.Br. or *Pandorea australis* (R.Br.) Spach).   Fig. 6.5–6.7, p. 27.

An extraordinarily variable climber. Leaves imparipinnate, leaflets in (2–)3–4 pairs, linear to lanceolate, ovate or elliptic, 2–10 × 0.5–4.5 cm long, entire or toothed. Flowers in terminal panicles, 5–13 cm long and some axillary cymes lower down, smaller than in previous species; calyx cupular, 2–3 mm long, undulate; corolla cream, streaked with dark purple or maroon especially on lower side, shortly infundibuliform, 1.5–2(–3.5) cm long, the lobes usually small and rounded, 2–5 × 2–4 mm. Fruit ellipsoid, (5–)9–12 × (1.7–)2.5–3 cm, rugose; seeds 2.5–3 × 1.5–2 cm including wing.

Native of Australia, New Guinea and Pacific Islands. Cultivated in Kenya: Nairobi, Dec. 1962, *Beecher* s.n.; Nairobi, Karen, Hort. Merryweathers, 1 Jan. 1967, *Gardner* in EA 13685.

FIG. 1. *MILLINGTONIA HORTENSIS* — **1,** inflorescence × ²⁄₃; **2,** leaf × ²⁄₃; **3,** capsule × ²⁄₃; **4,** seed × ²⁄₃. *PARMENTIERA ACULEATA* — **5,** inflorescence × ²⁄₃; **6,** leaf × ²⁄₃; **7,** capsule × ²⁄₃. 1, 2, from *Dening* H.4447/61; 3, 4, from *Dale* 2024; 5, 6 from *Hinton et al.* 7126; 6a from *Greenway* 2540; 7 from *Chandler* 1238. Drawn by Emmanuel Papadopoulos.

Parmentiera aculeata (*Kunth*) *Seem.* (*P. cerifera* sensu Brenan, T.T.C.L.: 72 (1949); *P. edulis* DC.). Fig. 1.5–1.7, p. 13.

Tree 5–10 m tall, usually with a stout thorn subtending each leaf but these sometimes lacking. Leaves often ± fasciculate, 3(–4–5)-foliolate or a few simple, leaflets elliptic-obovate, 1.5–6 × 0.6–3 cm with petiole narrowly winged, 1.3–3.5 cm long. Flowers solitary or in few-flowered fascicles, terminal or axillary near tops of old branches; calyx basically spathaceous, 2.5–4 cm long; corolla greenish white, campanulate, 5–7 cm long, with a transverse fold across lower side of the throat. Fruit linear, (5–)11–17 cm long and 1.2–3 cm in diameter, with thick ribs; seeds small, thin, 3–4 × 3 mm.

Native of Mexico to Honduras. Cultivated in Uganda: Entebbe, Sept. 1924, *Maitland* 125; Entebbe, area adjacent to Botanic Garden, May 1935, *Chandler* 1238; Kenya: Kisumu, 29 Dec. 1950, *Gamlen* H7/51 & 18 Jan. 1951, *Gamlen* H9/51 & 28 Feb. 1951, *Gamlen* H41/51; and Tanzania: Amani, Kiumba Plantation 2, 20 Oct. 1930, *Greenway* 2540; Dar es Salaam, Red Cross UWT/Upanga road junction, 23 Mar. 1993, *Mwasumbi* 16779.

Parmentiera cerifera *Seem.* Candle Tree (Dale, List Introd. Trees Uganda: 53 (1953))

Tree to 20 m, not aculeate. Leaves 3-foliolate; petiole conspicuously winged, 2.4–6.2 × ± 0.4 cm; leaflets elliptic to elliptic-rhomboid, 3–9.5 × 1.4–4 cm. Flowers solitary or 2–3 in fascicles on abbreviated shoots on trunk or main branches; calyx spathaceous, 2–5 cm long; corolla white or greenish white, 3.7–6.4 cm long, with a transverse fold across the throat. Fruit waxy yellow, linear, 39–54 × 1–2.4 cm.

Native of Panama. Cultivated in Uganda (Entebbe Botanic Gardens, fide Dale).

Pithecoctenium crucigerum (*L.*) *Gentry*, Monkey's Hairbrush

Climber. Leaves 2–3-foliolate, the terminal one often replaced by a tendril; leaflets ovate or ± round, 3.3–18 × 2–15 cm, cordate at base, acuminate at apex; petioles 3–7 cm long. Inflorescences terminal, up to 15-flowered; calyx 8–12 mm long; corolla white with yellow throat, tubular-campanulate, 3.5–6 cm long often with 90° bend in middle of tube. Fruit oblong to ellipsoid, 24–31 × 5–7.5 cm, echinate; seeds oblong, ± 7 × 2.5 cm including the extensive hyaline wings.

Native of Mexico to Northern Argentina and Uruguay. Has been grown in Kenya: Nairobi, Karen, 11 Jan. 1973, *R.R. Scott* in EA 15244.

Podranea ricasoliana (*Tanfani*) *Sprague*, Port St. John's Creeper, Zimbabwe creeper (T.T.C.L.: 72 (1949) as *P. brycei* (N. E. Br.) Sprague; Jex-Blake: 141 as *Tecoma brycei* N. E. Br. and *T. mackenii* W. Wats.). Fig. 2.7–2.10, p. 15.

Strong-growing climbing shrub to 10 m or more. Leaves imparipinnate with 3–5(–7) pairs of leaflets; tendrils absent; leaflets narrowly ovate to ovate-lanceolate, 3.5–9 × 1.3–2.6 cm. Flowers in few- to many-flowered panicles up to 40 cm long, sweet-scented; calyx campanulate, 0.7–1.5(–2.2) cm long; corolla pale pink or rose with darker red streaks inside, campanulate, 3.5–5(–6.5) cm long, 2–3.5(–5) cm wide at throat, narrowing abruptly to 3–5 mm wide at base; lobes ± round, 1.3–2.8 cm long. Fruit linear, 30–45 × ± 1.5 cm; seeds 0.6–1 × ± 3 cm including wing.

Native of Zimbabwe, Malawi, Mozambique and South Africa; has been widely cultivated in East Africa and it is not clear if it may actually have become naturalised in places. Kenya: Naivasha District: Lake Naivasha, 5 May 1967, *S.F. Polhill* 217; Nairobi Arboretum, 25 Feb. 1952, *G.R. Williams* 350; Nairobi, Loreto ridge, 29 Oct. 1973, *Gillett* 21692; and Tanzania: Masai District; Ngorongoro, 18 May 1982, *Chuwa & Bararea* 2428K; Lushoto District: Amani, 4 July 1950, *Verdcourt* 282; Iringa District: Chiwanjo, 18 Aug 1933, *Greenway* 3574 & Mufindi, Lugoda Estate, 1 April 1988, *Bidgood et al* 872.

Fig. 2. *SARITAEA MAGNIFICA* — **1,** inflorescence × ²/₃; **2,** leaves × ²/₃. *PYROSTEGIA VENUSTA* — **3,** inflorescence × ²/₃; **4,** leaf × ²/₃; **5,** capsule × ²/₃; **6,** seed × ²/₃. *PODRANEA RICASOLIANA* — **7,** inflorescence × ²/₃; **8,** leaf × ²/₃; **9,** capsule × ²/₃; **10,** seed × ²/₃. 1 from *Biegel* 2845; 2 from *Greenway* 6835; 3, 4, from *Stewart* 766; 5, 6 from *Gardner* 1768; 7–10 from *Ryding* 1166. Drawn by Emmanuel Papadopoulos.

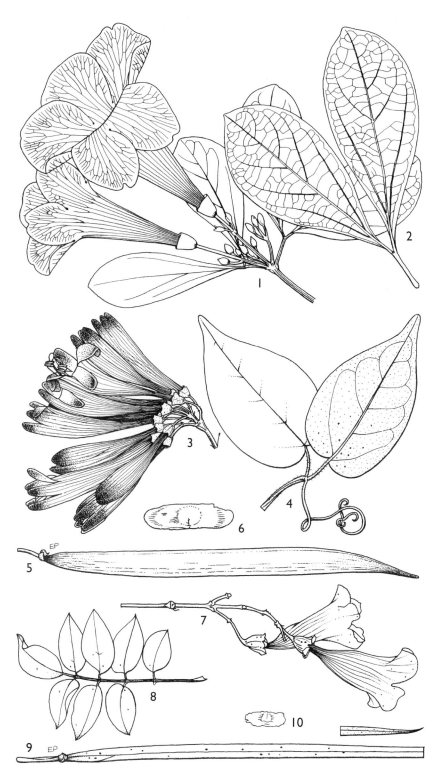

NOTE. *P. brycei* has usually been kept separate from *P. ricasoliana* (e.g. Diniz in F.Z. 8(3): 79(1988)). Supposedly restricted to South Africa but none of the characters, flower size, exsertion of narrow part of corolla tube and presence or absence of hairs in corolla tube actually holds up. Sandwith long ago on a determination label suggested they were probably the same and it is clear Gentry considered them conspecific.

Pyrostegia venusta (*Ker-Gawl.*) *Miers** Golden shower (T.T.C.L.:72 (1949); Jex-Blake: 132, 332, 354 as *Bignonia venusta*).   Fig. 2.3–2.6, p. 15.

Extensive woody climber reaching tops of tall trees or covering houses. Leaves with 2 leaflets, the terminal replaced by a tendril; leaflets ovate to elliptic, 3.5–8 × 2.5–6 cm, acute to acuminate at the apex. Flowers in terminal or axillary subumbelliform panicles; calyx campanulate (3–)4–7 mm long with very short teeth ± 0.5 mm long; corolla brilliant waxy orange, yellow or red, tubular-funnel-shaped, (4–)4.5–8.5 cm long, 0.7–1.2 cm wide at the throat, lobes oblong, 1–2.5 × 0.4–0.7 cm. Fruits flat, linear, 16–33 × 1.2–1.6 cm; seeds ± 0.9 × 1 cm, slightly bilobed with brownish hyaline wings.

Native of Brazil; this is one of the most widely planted showy climbers in the tropics and must be in almost every sizeable garden in E Africa; being so well known there are not many collections. Kenya: Nairobi, Hort. Stewart, July 1963, *J. Stewart* 766 and Tanzania: Moshi District: Old Moshi, 11 May 1961, *Machangu* 61; Lushoto District: Amani, Boma House, 26 Aug. 1929, *Greenway* 1691 & Lushoto, near silviculturist's office, 23 July 1970, *Mshana* 89 & Derema Tea Estate, 28 June 1997, *Hizza & Mndolwa* 18.

Radermachera sinica (*Hance*) *Hemsl.*

Tree to 7 m tall. Leaves 3-pinnate, leaflets narrowly ovate, to 7 × 3 cm, terminal up to 8.5 cm. Flowers in narrow few-flowered terminal panicles. Calyx 1.8–4 cm long; corolla (7.2–)8–13.5(–15) cm long, tube infundibuliform, 6–11 cm long, lobes elliptic or broadly so, to 4 × 3 cm, with crenulate margin. Fruit smooth, linear, cylindrical, often twisted, 35–65 × 1–1.5 cm.

Native of South East Asia, China, Taiwan. No material has been seen from East Africa, but the species has been cultivated in Zimbabwe.

Radermachera xylocarpa (*Roxb.*) *K.Schum.*

Tree to 20 m tall. Leaves 2-pinnate, leaflets ovate, to 12 × 6.5 cm. Flowers in wide many-flowered terminal panicles. Calyx 0.7–1.5 cm long; corolla 3.5–5.5(–7) cm long, tube campanulate, 2.5–4 cm long, lobes elliptic or broadly so, to 3 × 2 cm. Fruit verrucose, linear, cylindrical, woody, ± 55 × 2 cm.

Native of India. No material has been seen from East Africa, but the species has been cultivated in Zimbabwe.

Saritaea magnifica (*Steenis*) *Dugand* (Jex-Blake: 132, 332 as *"Bignonia magnifica"*** and *Arrabidaea magnifica* Steenis).   Fig. 2.1–2.2, p. 15.

Climber to 10 m. Leaves with two leaflets and usually a tendril; leaflets obovate, 3.5–10.5 × 2–7 cm, obtuse; pseudostipules leafy, 0.6–4.2 × 0.4–2 cm. Flowers in basically terminal few-flowered panicles; calyx 6–10 mm long; corolla magenta-pink or purple, white in throat with purple-red nectaries (Ruffo reported flowers blue), funnel-shaped, tube ± 6.5 cm long, 1.5–2 cm wide at the throat; lobes ± 3 × 2.5–3 cm. Fruit linear, flattened, ± 29 × 1 cm; seeds ± 3.5 × 0.8 cm including wing.

Native of Colombia and Ecuador; has been grown in E Africa, all material seen is from Tanzania: Lushoto District: Amani Nursery, 4 Feb. 1939, *Greenway* 5856 & Amani, 28 Oct. 1969, *Ruffo* 273 & 10 Feb. 1971, *Furuya* 135 & Amani, Bustani, 31 Oct. 1969, *Ngoundai* 421; Zanzibar, 1927, *Toms* 9M.

---

* Sometimes given as (Ker-Gawl.) Baill. but Miers definitely made the combination.
** Not *Bignonia magnifica* Bull; see discussion by van Steenis in Bull. Jard. Bot. Buitenzorg III, 10: 193 (1928).

Tabebuia aurea (*Manso*) *S.Moore* (*T. argentea* (Bur. & K.Schum.) Britton)
Tree to 16 m tall with thick corky bark. Leaves palmately 5–7-foliolate, glabrous; leaflets narrowly oblong–oblanceolate to oblong-elliptic, 13–14 × 3.5–9.5 cm, glabrous, rounded to retuse at the apex, rounded to subcordate at the base; petiole up to 14 cm long; petiolules 3.5–5 cm long. Flowers in terminal panicles, showy; calyx campanulate, irregularly bilabiate, 8–16 mm long; corolla bright yellow, tubular-funnel-shaped, 4.2–6.6 cm long, 1.2–2.5 cm wide at throat; lobes 1.2–2.2 cm long. Fruit oblong, 8.5–15 × 1.7–3 cm; seeds 2-winged, 4.5–5.5 × 2 cm.
Native of South America. Has been cultivated in Tanzania: Lushoto District: Amani Chini, 2 Nov. 1944, *Fernie* s.n.

Tabebuia chrysantha (*Jacq.*) *Nicholson* subsp. **pluvicola** *Gentry*
Tree to 30 m tall with usually rather smooth bark but Kenya material reported as small tree with rough brown bark. Leaves palmately 5(–7)-foliolate, leaflets elliptic to oblong-obovate, up to 25 × 14 cm, acute to shortly acuminate at the apex, obtuse to truncate at the base, usually with stellate pubescence on main venation above and more extensive beneath; petiole 6–30 cm long, stellate-pubescent; petiolules 3–8 cm long. Flowers ± precocious, in contracted terminal panicles, stellate-rufescent; calyx campanulate, 12–19 mm long; corolla golden yellow, tubular to funnel-shaped; tube 4–8 cm long, 1.8–3 cm wide at throat; lobes 1.5–3 cm long. Fruit linear–cylindrical, 30–90 × 1.5–2.4 cm, glabrescent; seeds 2-winged 2.5–3.4 × 0.6–0.9 cm.
Native of Costa Rica. Has been cultivated in Kenya: Kiambu District, Muguga, 6 Dec. 1974, *Dyson* 678.
NOTE. R.O. Williams (U.O.P.Z.: 458) mentions "*Tabebuia rufescens* (*Tecoma glomerata*)" as growing in Zanzibar gardens opposite the hospital and a native of Trinidad with golden flowers. Gentry (Fl. Neotrop. 25(2): 165 (1992)) gives *T. rufescens* J.R. Johnston as a synonym of *T. chrysantha* subsp. *chrysantha* and Venezuela as the type locality. *Tecoma glomerata* I have not been able to trace. It seems likely that the tree introduced into Zanzibar was typical *Tabebuia chrysantha* but without material it is not certain. It may not now still be growing in the island.

Tabebuia chrysotricha (*Mart.*) *Standley*
Shrub or small tree 2–10 m tall, the branches rufous stellate-pubescent when young. Leaves palmately (3–)5-foliolate; leaflets oblong-obovate to oblong-elliptic, (1.5–)2–11(–15) × (1–)2–5.5(–9) cm, rounded to abruptly cuspidate-acuminate at the apex, obtuse to truncate at the base, stellate-pubescent; petiole 1–2.5 cm long; petiolules 0.2–8 cm long. Flowers in few-flowered subsessile terminal clusters; calyx tubular, 1–2 cm long, hairy; corolla yellow with some reddish lines at throat, tubular to funnel-shaped; tube 3.5–5.5 cm long, 1.5–3 cm wide at throat; lobes 0.5–1.5 cm long. Fruit linear-cylindrical, 11–38 × 0.8–1.2 cm, reddish or golden hairy; seeds 1.7–2.9 × 0.6–0.9 cm.
Native of Brazil and Argentina. Has been grown in Kenya: Kwale District, Diani, Robinson Club Boabab Hotel, 7 May 1986, *Robertson* 4156.

Tabebuia guayacan (*Seem.*) *Hemsley* (Jex-Blake: 127 as *Tabebuia "guaiacum"*)
Tree to 50 m, the bark with vertical ridges, wood extremely hard. Leaves palmately 5–7-foliolate; leaflets lanceolate to ovate, 9–30 × 3.7–13.5 cm, acuminate at apex, rounded to cuneate at the base, stellate-pubescent in axils of lateral nerves beneath; petiole 7–23 cm long, glabrous. Flowers in terminal panicles; calyx 7–15 mm long, stellate-pubescent; corolla yellow with reddish lines in the throat, tubular to funnel-shaped; tube 3.5–5.7 cm long, 1.2–2.2 cm wide at throat, lobes 2.2–3.8 cm long. Fruit linear-cylindrical, 29–61 × 1–2.9 cm, glabrous or stellate–pubescent; seeds 3.5–4 × 0.9–1.1 cm.
Native of Mexico to Venezuela and Peru. Has been reported from Uganda, Entebbe Botanic Gardens and also from Kenya, Nairobi Arboretum but all this material has proved to be *T. impetiginosa*; since it may occur (see note) a description is given.

NOTE. A sterile specimen from Kenya (Kwale District: Mazeras–Gandini, 22 April 1992, *Luke* 3095) might be *T. guayacan* as it was originally named. The thin young leaves have dried black and are pubescent above, more densely so beneath, the leaflets are long-acuminate. More adequate flowering material is needed.

Tabebuia heterophylla (*DC.*) *Britton.* White Cedar (*Tecoma leucoxylon* sensu U.O.P.Z.; 463 (1949)

Fast growing shrub or small to large tree up to 20 m tall or more. Leaves glabrous, palmately 3–5-foliolate; leaflets very variable, the terminal leaflet obovate to obovate-elliptic, the laterals elliptic to oblong-elliptic, 1–16 × 0.4–7.5 cm, obtuse to rounded at the apex and base; petiole 0.5–8(–14) cm long; petioles 0.2–5.5 cm long. Flowers in few to many-flowered panicles; calyx cupular, 7–12 mm long; corolla lavender, pale magenta or mauve with yellow throat, tubular to funnel-shaped, tube 3–5.5 cm long and lobes 0.8–2 cm long, glabrous outside. Fruit linear-cylindrical, 7–20 × 0.6–1 cm; seeds 2-winged, 20–30 × 7–9 mm.

Native of West Indies where very widely distributed. Has been grown in Tanzania: Zanzibar, Kizimbani Experimental Station, fide U.O.P.Z.

NOTE. The complex nomenclatural tangles surrounding this species are explained at length by Sandwith in K.B. 1953: 453–454 (1954) and Gentry in Fl. Neotrop. 23 (2): 196–197 (1992). The reference to *Tabebuia pallida* in F.Z. Bignon.: 8(3): 62 (1988) refers to this species. True *T. pallida* (Lindl.) Miers has mainly 1-foliolate leaves sometimes with some 3-foliolate and is restricted to the Lesser Antilles. It is very close to *T. heterophylla* and Gentry (Fl. Neotrop. 25(2): 234 (1992)) doubts it is a true biological species.

Tabebuia impetiginosa (*DC.*) *Standley*

In East Africa usually a deciduous shrub to 6 m but typically a tree to 30 m tall with rather smooth bark, sometimes flowering when almost leafless. Leaves palmately 5(–7)-foliolate; leaflets ovate to elliptic, 5–19 × 1.5–9 cm, cuneate to almost subcordate at the base, entire or slightly serrate towards apex, acuminate at the apex, glabrous to slightly pubescent above, ± densely pubescent beneath; petiole 4–13 cm long ± pubescent; petiolules 1–4 cm long. Flowers in a terminal usually congested cluster or panicle; calyx 4–9 mm long, densely pale velvety-pubescent; corolla magenta or orange-pink, the throat yellow at first becoming purplish, base of tube sometimes orange, pubescent outside, narrowly funnel-shaped; tube 2.5–5 cm long, 1.2–2 cm wide at the throat; lobes 1–2 cm long. Fruit cylindrical, 12–56 × 1.3–2.6 cm, glabrous; seeds 3.4–8 × 1–1.6 cm.

Native of Central and South America. Uganda: Entebbe Botanic Gardens, *Asst. Agric. Officer* 2; Kenya: Nairobi Arboretum, Aug. 1945, *Bally* in CM 12360 & 15 Dec. 1952, *Greenway* 8746.

Tabebuia rosea (*Bertol.*) *DC.*, Rosy Trumpet tree (T.T.C.L.: 73 (1949) as *T. pentaphylla* auct.*; Dale, List Introd. Trees Uganda: 67 (1953) as *T. pentaphylla* and *T. rosea* not realising they were synonyms; Jex-Blake: 331 as *T. pentaphylla*).

Fig. 3.1–3.4, p. 19.

Tree 12–25(–30) m tall, bark narrowly vertically fissured with corky ridges, often flowering when leafless. Leaves palmately (3–)5-foliolate; leaflets elliptic to elliptic-oblong, 8–35 × 3–18 cm, acute to acuminate at the apex, cuneate to rounded at the base, glabrous; petiole 5–32 cm long; petiolules 0.2–3.5 cm long. Flowers many, in terminal subumbelliform panicles; pedicels up to 2 cm long; calyx 1–2 cm long, bilabiate, glabrous; corolla pink, magenta or sometimes white the throat yellow,

---

* *Tabebuia pentaphylla* (L.) Hemsley is a synonym of *T. heterophylla* (DC.) Britton but was long wrongly associated with the above species. See Sandwith in K.B. 8: 453–454 (1954) and Gentry, Fl. Neotropica 25 (II): 196–197 (1992) for explanation of this complicated tangle.

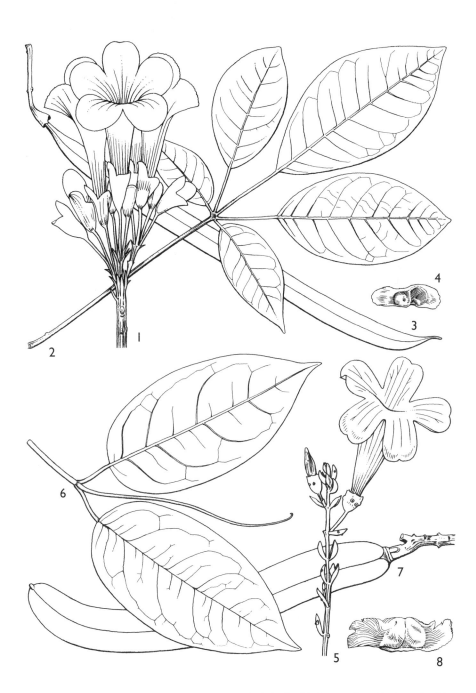

FIG. 3. *TABEBUIA ROSEA* — **1,** inflorescence × ²/₃; **2,** leaf × ²/₃; **3,** capsule × ²/₃; **4,** seed × ²/₃. *ADENOCALYMMA MARGINATUM* — **5,** inflorescence × ²/₃; **6,** leaf × ²/₃; **7,** capsule × ²/₃; **8,** seed × ²/₃. 1 from *Gardner* EA 14458; 2–4 from *Hinton* 10022; 5 from *Robertson* 5090; 6 from cult. RBG Kew H. 3325/66; 7, 8 from *Klein* 5737. Drawn by Emmanuel Papadopoulos.

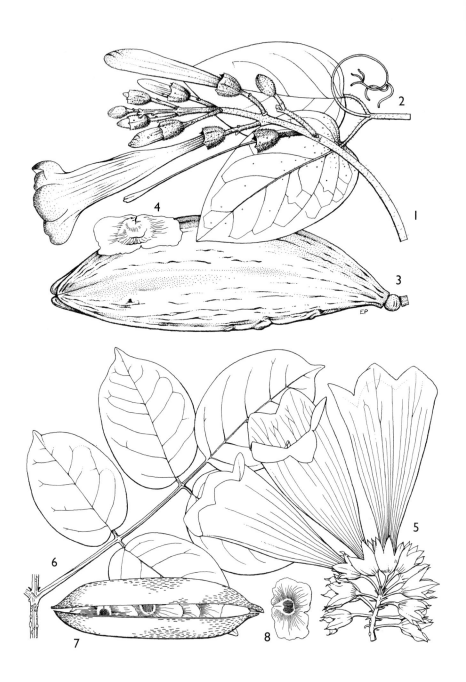

F<small>IG</small>. 4. *DISTICTIS BUCCINATORIUS* — **1,** inflorescence × ²/₃; **2,** leaf × ²/₃; **3,** capsule × ²/₃; **4,** seed × ²/₃. *TECOMANTHE DENDROPHILA* — **5,** inflorescence × ²/₃; **6,** leaf × ²/₃; **7,** capsule × ²/₃; **8,** seed × ²/₃. 1, 2 from *Greenway* 10900; 3, 4, from *Varnaro Pujira* 37-1903; 5 from *Ng'ang'a* EA 16210; 6 from *Pattison* 2158; 7, 8 from *Bartlett* H/2572/1911. Drawn by Emmanuel Papadopoulos.

narrowly funnel-shaped, glabrous outside, 5–8 cm long, tube 3–5.8 cm long and 1.5–3 cm wide at throat; lobes 2–2.5 cm long. Fruit linear-cylindrical, 22–38 × 0.9–1.5 cm; seeds 2.8–4.4 × 0.7–1 cm.

Native of West Indies, Central and Northern South America. Has been grown in Uganda: Entebbe Botanic Gardens, fide Dale; Entebbe, Lake Victoria Hotel grounds, 7 Sept. 1970, *H. Gordon* EA14458; Kenya: Pipeline Co. H.Q. Mombasa, 8 Aug. 1983, *Pipeline Co.* in EA 16951; Tanzania: Lushoto District: Mombo Arboretum, plot 15B, 13 Oct. 1971, *Ngonyani* 48; Amani, Plantation AC2, 24 June 1929, *Greenway* 1596 & Amani Nursery, 20 Mar. 1973, *Ruffo* 674; Zanzibar, Kisakasaka, 26 Nov. 1999, *Fakih* 516.

Tecoma includes naturalised taxa. These and the cultivated species are treated on p. 26.

Tecomanthe dendrophila (*Bl.*) *K.Schum.* (*T. venusta* S.Moore).
Fig. 4.5–4.8, p. 20.
Creeper or liane up to 20(–30) m. Leaves pinnate 3–5-foliolate, leaflets ovate to oblong-lanceolate, 3–13 × 1.5–7 cm, entire or notched at apex. Flowers 6–20 in condensed pendulous clusters borne on the old wood; calyx green, tinged red or purple, 1.2–4 cm long; corolla with tube pale to deep pink, narrowly funnel-shaped, 7–11 cm long including lobes 0.7–1.5 × 1–2 cm. Fruit linear, woody, 17–30 × 3–3.7 cm; seeds 2.5–3.5 × 1.2–1.5 cm.

Native to Moluccas, New Guinea and Solomon Islands. Kenya: Nairobi City Park, 10 Nov. 1976, *Ng'ang'a* EA 16210.

Tecomaria is indigenous. Native and cultivated taxa are treated on p. 24.

Tourettia lappacea (*L'Herit.*) *Willd.* (Jex-Blake: 142 as *T. volubilis* Gmel.)
Annual creeper. Leaves with two pinnae each pedately 5-foliolate or trifoliolate with a branched tendril between; leaflets ovate, 3–8 × 1.5–5 cm, acute, conspicuously toothed. Inflorescences subspicate, the pedicels very short with scarlet bracts; terminal flowers mostly sterile with bright red calyces; calyx of fertile flowers 1.3–1.5 cm long, deeply bipartite; corolla purple or green, tubular, 1.6–1.8 cm long, tube ± 1 cm long, the lobes fused to form a lip 6–8 mm long. Fruit ± woody, ovoid, 3–3.5 × 1.7 cm, densely covered with hooked bristles; seeds flat, 6–7 × 5–5 mm.

Native of Central America and Western South America. Has presumably been grown in Kenya but no material has been seen.

NATIVE & NATURALISED SPECIES

Newbouldia laevis (*P.Beauv.*) *Bur.*

Tree to 15 m tall with pinnate leaves, the leaflets 4-jugate, to 16.5 × 8.5 cm, leathery, glabrous, coarsely serrate. Flowers in a narrow terminal panicle; calyx spathaceous, ± 2 cm long; corolla pink, mauve or purple, 4–5 cm long. Fruit linear, ± 28 cm long with inrolled valves.

NOTE. A sheet (*Maitland* s.n., Uganda, Mbarara Region, 1925) labelled as this was thought by Gillett and Brenan (in sched.) to be a mislabelled duplicate of *Maitland* 282 or 772 from the Cameroons. Presumably the sheet labelled Uganda was thought to have been wild material rather than cultivated. The nearest confirmed records are from Cameroon, Gabon and Western Congo, but disjunctions of this magnitude are not unheard of and it is just possible that the species might occur in Uganda.

1. MACFADYENA

A.DC. in DC., Prodr. 9: 179 (1845)

Doxantha Miers in Proc. Hort. Soc. 3: 188 (1863)

Climber. Branches slender with glandular patches between the nodes. Leaves with 2 leaflets, the third replaced with a trifid tendril with claw-like divisions; pseudostipules small, subulate. Flowers in axillary cymes, sometimes only 2–3 or solitary. Calyx truncate to spathaceous, split to middle, bilabiate or irregularly lobed. Corolla tubular-campanulate, glabrous outside. Anthers glabrous. Ovary oblong, scaly, puberulous or ± glabrous; ovules 2–4-seriate. Disk annular. Fruit a linear capsule. Seeds 2–winged.

Although about 35 species are listed, Gentry comments that there are only 3–4 good species in South and Central America and the West Indies, one very commonly and widely cultivated and apparently becoming naturalised in East Africa.

Macfadyena unguis-cati (*L.*) *Gentry* in Brittonia 25: 236 (1973); Liben in F.A.C., Bignon.: 36 (1977); Bidgood in F.Z. 8(3): 64 (1988). Lectotype: Tab. 94 in Plumier, Pl. Amer. (1756), chosen by Nasir, Fl. Pakistan 131: 18 (1979)

Climber to 10 m, often forming dense covering on trunks etc. Roots sometimes with ellipsoid tubers. Leaflets elliptic, 5–16 × 0.9–7 cm (smaller when young), narrowed to a ± rounded base, acutely acuminate at the apex, glabrous; petiole 1–4.7 cm long, petiolules 0.5–2.5 cm long. Inflorescences 1–3(–15)-flowered. Calyx 0.5–1.8 cm long with undulate margin. Corolla bright yellow, sometimes with orange lines in throat, 4.5–10 cm long; tube 3.3–6.9 cm long, 1.2–2.5 cm wide at throat; lobes 1.3–3.1 cm long. Capsule 26–95 × 1–1.9 cm, narrowed at apex. Seeds thin, strongly winged at both ends, narrowly rectangular, ± 4.8 × 0.8 cm. Fig. 5.1–5.5, p. 23.

UGANDA. Mengo District: W Mengo, Kyebando, 14 Mar. 1990, *Rwaburindore* 2959!
KENYA. Nairobi District: Nairobi, Riverside Drive, Jan. 1953, *Bally* 8487!; Kisumu–Londiani District: Fort Ternan, Kipchui Farm, July 1958, *Firth* H226/58!; Kwale District: Shimba Hills, Baraza Park, 29 July 1993, *Luke* 3610!
TANZANIA. Lushoto District: Amani Nursery, Feb. 1978, *Ruffo* 1498! & roadside near Amani Nursery, 30 Sept. 1977, *Sigara* 10!
DISTR. **U** 4; **K** 4, 5, 7; **T** 3; Central and South America, West Indies; widely cultivated in tropics and greenhouses
HAB. Coffee shambas, *Erythrophleum, Bombax, Albizia* parkland; 350–400 m (naturalized specimens)

SYN. *Bignonia unguis-cati* L., Sp. Pl.: 623 (1753)
   B. tweediana Lindl. in Bot. Reg. 26: t. 45 (1840); Jex-Blake, Gard. E. Afr., ed. 4: 132, 332, 356 (1957). Type: Argentina, imported from Buenos Ayres by *Strangeways* (CGE?, holo.)
   Doxantha unguis-cati (L.) Miers in Proc. Hort. Soc. 3: 190 (1863) (as '*D. unguis*'*)

NOTE. At least some of the specimens cited appear to form naturalised populations but this needs confirmation. Rwaburindore states the tendrils are sticky.

---

* It has been argued that this combination is not valid and should be attributed to Rehder but I think it is a correctable error and should be allowed to stand.

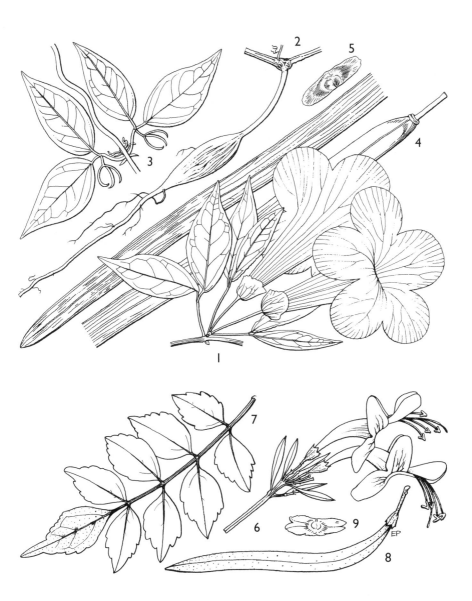

FIG. 5. *MACFADYENA UNGUIS-CATI* — **1,** habit of shoot with flowers × $\frac{1}{2}$; **2,** roots with tuber; **3,** leaves × $\frac{1}{2}$; **4,** capsule × $\frac{1}{2}$; **5,** seed × $\frac{1}{2}$. *TECOMARIA CAPENSIS* — **6,** shoot with flowers × $\frac{2}{3}$; **7,** leaf × $\frac{2}{3}$; **8,** capsule × $\frac{2}{3}$; **9,** seed × $\frac{2}{3}$. 1 from *Mackay* s.n.; 2, 4, 5 from *Firth* H226/58; 3 from *Biegel* 4725; 6–9 from *Friis et al.* 9313. 1–5, drawn by Eleanor Catherine, 6–9 drawn by Emmanuel Papadopoulos.

2. TECOMARIA

(Endl.) Spach, Hist. Nat. Veg. Phan. 9: 137 (1840)

Shrubs or small trees with imparipinnate leaves. Flowers in terminal racemes or panicles. Calyx campanulate, 5-lobed. Corolla with curved very narrowly funnel-shaped tube and bilabiate limb, the upper lobe 3-fid, the lower 2-fid. Stamens well-exserted, didynamous; thecae connate for the upper third, slightly diverging below. Disk cupular; ovary bilocular with ovules numerous, 4-seriate in each locule. Fruit a linear-oblong compressed dehiscent capsule. Seeds with membranous wings.

A single species in tropical and S Africa but much cultivated elsewhere as an ornamental. Only one subspecies occurs wild in East Africa but the other is cultivated. K. Schumann (E.&.P., Pf. 4 (3b): 229 (1895)) kept *Tecoma* and *Tecomaria* separate but retained three South and Central American species in *Tecomaria* presumably because they had exserted stamens. The two genera are very close with most species of *Tecoma* having the stamens included and all have much more regular straight corollas; the anthers are not in any way connate. This is explained away by those who unite the genera as pollination adaptations but we do not see why being able to explain a difference in characters means they are of no value. Gentry at first kept them separate but later combined them (Fl. Neotropica 25(2): 273–4 (1992)).

Tecomaria capensis (*Thunb.*) *Spach*, Hist. Nat. Veg. Phan. 9: 137 (1840); Seem. in J.B. 1: 21 (1863); Sprague in F.T.A. 4(2): 514 (1906); Jex-Blake, Gard. E. Afr., ed. 4: 128, 356 (1957); F.F.N.R: 380 (1962): Palmer & Pitman, Trees S. Afr. 3: 2001 (1973); Brummitt in B.J.B.B. 44: 421 (1974); Liben, F.A.C., Bignon.: 11 (1977); Gentry in Fl. Gabon 27: 54, t.11 fig. 4 (1985); Diniz in F.Z. 8(3): 64, t.10 (1988) & in C.F.A., Bignon.: 13 (1993). Type : South Africa, without locality, *Thunberg* (UPS, holo.; microfiche!)

Shrub or small tree, sometimes somewhat scrambling, 1.5–7(–10) m tall, with pale brown bark; branches copiously lenticellate. Leaves 10–25(–33) cm long; petiole 2–7(–9) cm long; leaflets in 2–6 pairs, elliptic, ovate or ± round, 1.7–4.8(–8.5) × 1.3–3(–3.8) cm, rounded to cuneate at the base, crenate to toothed at margin, acute to rounded at the apex with the tip acuminate to mucronate or rarely retuse, glabrous or glabrescent above, glabrous beneath save base of midrib and nerve-axils; petiolules absent or up to 1.2 cm long. Inflorescences 10–32 cm long; peduncles 7–20 cm long; bracts mostly linear, 3–7 mm long; pedicels 4–11(–13) mm long. Calyx 4–20(–23) mm long, lobed to ± halfway; lobes broadly triangular. Corolla orange to scarlet or rarely yellow; tube 2–3.5 cm long, 5-lobed, lobes 1–1.5 cm long, upper 3 joined above half way, lower 2 free, recurved. Anthers 3–5 mm long. Ovary oblong, ± 4 mm long. Fruit 4.5–13 (?–19.5) cm long, 1.1–1.4 cm wide, attenuate at both ends. Seeds ± rectangular in outline including the flat hyaline membranous wing, 1.3–2.1 cm long, 6–10 mm wide. Fig. 5.6–5.9, p. 23.

Syn. *Bignonia capensis* Thunb., Prodr. Pl. Cap. 2: 105 (1800)
    Tecoma capensis (Thunb.) Lindl. in Bot. Reg. 12: t. 1117 (1827); Jex-Blake, Gard. E. Afr., ed. 4: 141 (1957); Gentry in Fl. Neotropica 25(2): 277 (1992)

1. Calyx (4–)5–8(–10) mm long, campanulate to
   shortly cylindrical; leaves with 2–3(–5) pairs of
   leaflets (only cultivated in Flora area) 2
   Calyx (8–)10–18(–23) mm long, cylindrical; leaves
   with (3–)4–5(–6) pairs of leaflets . . . .   subsp. *nyassae*
2. Corolla orange or scarlet . . . .   subsp. *capensis* var. *capensis*
   Corolla yellow . . . .   subsp. *capensis* var. *flava*

subsp. **nyassae** (*Oliv.*) *Brummitt* in B.J.B.B. 44: 421 (1974); Liben, F.A.C., Bignon.: 11, t. 2.C–D & t. 3 (1977); Diniz in F.Z. 8(3): 66, t.10 (1988) & C.F.A. 122, Bignon.: 13, t.1 (1993). Type: Tanzania, "lower plateau N of Lake Nyassa", *Thomson* s.n. (K!, holo.!)

Shrub or small tree to 7(–10) m tall. Leaflets in (3–)4–5(–6) pairs and often larger than in typical subspecies. Calyx (8–)10–18(–23) mm long with lobes 3–6 mm long or more. Corolla red or orange.

TANZANIA. Lushoto District: Lushoto, 1 Sept. 1912, *Semsei* 3519!; Iringa District: about 192 km S of Iringa, just S of Sao Hill, 27 July 1959, *Verdcourt* 2338!; Songea District: Matengo Hills, Lupembe Hill, 27 May 1956, *Milne-Redhead & Taylor* 10530!
DISTR. **T** 3–8; Congo-Kinshasa, Angola, Zambia, Malawi, Mozambique
HAB. Upland grassland and scattered tree grassland, rain forest and mist forest probably always on margins, rocky places in *Brachystegia* woodland, roadsides in grassland areas; (1200–)1350–2750 m

SYN. *Tecoma nyassae* Oliv. in Hook. Ic. Pl. 14: t. 1351 (1881); Gentry, Fl. Neotropica, 25(2): 282 (1992)
    T. shirensis Bak. in K.B. 1894: 30 (1894). Type: Malawi, Shire Highlands, *Buchanan* 219 (K!, lecto.; BM, isolecto., chosen by Brummitt)
    Tecomaria nyassae (Oliv.) K.Schum. in E. & P., Pf. 4 (3B): 230 (1895); Sprague in F.T.A. 4(2): 515 (1906); T.T.C.L.: 73 (1949); Brenan in Mem. N.Y. Bot. Gard. 9: 18 (1954)
    T. shirensis (Bak.) K.Schum. in P.O.A. C: 36 (1895); Sprague in F.T.A. 4. (2): 515 (1906); T.T.C.L.: 73 (1949)
    Tecoma whytei C.H.Wright in K.B. 1897: 275 (1897). Type: Malawi, Zomba Plateau, *Whyte* s.n. (K!, holo.)
    T. nyikensis Bak. in K.B. 1898: 159 (1898). Type: Malawi, Nyika Plateau, *Whyte* 112 (K!, holo.)
    Tecomaria rupium Bullock in K.B. 1931: 274 (1931); T.T.C.L.: 73 (1949). Type: Tanzania, Dodoma District, N Mpwapwa, *Greenway* 2425 (K!, holo.; EA!, iso.)

NOTE. Gentry retains *T. nyassae* and *T. capensis* as separate species stating they are easily distinguishable vegetatively. Although this is often so with the latter having leaflets with more obvious more rounded teeth, examination of hundreds of specimens shows this is not a constant character.

    T. capensis (Thunb.) Spach subsp. *capensis*, 'Cape Honeysuckle', is widely cultivated in East Africa and throughout the Tropics. Apart from the normal orange or red-flowered plant a pure yellow variety occurs in East Africa which has not been reported from anywhere else save on a determination label by Sandwith mentioning "a similar form is in cultivation in a garden on the French Riviera". It has been given a name for the convenience of horticulturalists*. Var. *capensis* is widely cultivated e.g. Uganda, Mengo District, Makerere University Botany Garden, 5 May 1971, *Lye* 6029!; Kenya, Nakuru District: Njoro, Plant Breeding Station, 20 June 1976, *Gitonga* 115!; Kiambu District: Muguga, Hort. Greenway, 10 Feb. 1963, *Greenway* 10866!; & Karen, Hort. Gardner, 12. Nov. 1965, *Gardner* in EA 13398!; Tanzania, Lushoto District: Lushoto Township, 8 Dec. 1959, *Willan* 470 & Boma road, 13 Oct. 1970, *Mshana* 124 & Mang'ula, near Pentecoste Church, 7 Apr. 1982, *Kisena & Mtui* 12.

    Var. *flava* has been seen mostly from around Nairobi: Mar. 1953, *Verdcourt* 918; Nairobi Arboretum, 27 Apr. 1953, *G.R. Williams* 541 & Nairobi, Closeburn Nursery, 6 May 1953, *Greenway* 8770; Tanzania, Lushoto Hotel, *Shabani* 925. The plant is not listed in Grahame Bell's Closeburn Catalogue (1949) so must have appeared in Nairobi in the early 1950s whether a local mutant or brought in from elsewhere is not known. Subsp. *nyassae* has also been cultivated but much less frequently. Kenya, Nairobi, Scott Agricultural Labs., Mar. 1960, *Verdcourt* 2624 (peculiar fasciated form) & Kiambu District, Muguga, Hort Greenway 13 Feb. 1963, *Greenway* 10867 & **K** 3 (? District) Kaptel (not traced), 29 Oct. 1985, *Ekkens* 324. Tanzania, Lushoto Arboretum, 25 Aug. 1971, *Issa* 96.

---

* **Tecomaria capensis** (*Thunb.*) *Spach* subsp. **capensis** var. **flava** Verdc. **var. nov.** a var. *capensi* corolla pure flava differt. Typus, Kenya, Kiambu District, Muguga, Hort. Greenway, 15 Feb. 1963, *Greenway* 10868 (K!, holo.; EA, iso.)

3. **TECOMA**

Juss., Gen. Pl.: 139 (1789); Gentry in Fl. Neotropica 25(2): 273–293 (1992)

Shrubs or small trees with simple, 3-foliolate or imparipinnate leaves; leaflets serrate. Flowers in terminal racemes or panicles. Calyx cupular, shallowly 5-lobed. Corolla yellow or orange-red, tubular to campanulate, glabrous outside. Stamens included, just reaching throat or exserted; anthers divaricate to basally divergent, glabrous or pilose. Ovary narrowly cylindrical, the ovules 2-seriate in each locule. Disk cylindrical. Fruit a linear capsule, somewhat compressed, dehiscing perpendicular to the septum. Seeds thin with 2 hyaline membranous wings.

A genus of 12 species in the tropics and subtropics of the New World. One species is very widely cultivated and has become naturalised. Several others are also cultivated in East Africa.

1. Leaflets very deeply toothed . . . .   *T. stans* var. *angustata*
   Leaflets serrate 2
2. Leaflets densely woolly pubescent to velvety beneath . . .   *T. stans* var. *velutina*
   Leaflets glabrous or only sparsely pubescent 3
3. Upper part of leaf-rhachis narrowly but distinctly
   winged; corolla narrowly funnel-shaped with stamens
   well included . . . .   *T. smithii*
   Leaf-rhachis not winged; stamens included or anthers
   just exserted 4
4. Corolla very narrowly funnel-shaped; anthers just
   exserted from tube . . . .   *T. tenuiflora*
   Corolla campanulate-funnel-shaped; stamens entirely
   included in tube . . . .   *T. stans* var. *stans*

Tecoma stans (*L.*) *Kunth*, Nov. Gen. Pl. 3: 144 (1819); T.T.C.L.; 73 (1949); U.O.P.Z.: 463 (1949); Dale, List. Introd. Trees. Ug.: 67 (1953); Jex-Blake, Gard. E. Afr. ed. 4: 128, 354, 356 (1957); Liben, F.A.C., Bignon.: 8, t. 3.A–B (1977); Gentry in Fl. Gabon 27: 53 (1985); Blundell, Wild Fl. E. Afr.: 382, t. 397 (1987); Bidgood in F.Z. 8(3): 61 (1988); Gentry in Fl. Neotrop, 25(2): 285 (1992); Tardelli & Settesoldi in Fl. Som. 3: 306 (2006). Lectoype: Tab. 54 in Plumier, Pl. Amer. (1756), chosen by Gentry

Shrub or small tree, 2.5–10 m tall, with dark ridged bark, and usually distinctly lenticellate branchlets. Leaves 3–9-foliolate or first pair of leaves on a branch sometimes simple; leaflets ovate to lanceolate, 2.5–15 × 0.8–6 cm, cuneate at the base, serrate (in one variety deeply so), acute to acuminate at the apex, glabrous, pubescent or velvety beneath; petiole 1–9 cm long. Flowers up to 20, in terminal or subterminal racemes. Calyx elongate-cupulate, 3–7 mm long, 5-toothed, the teeth triangular, ± 1 mm long. Corolla bright yellow or sometimes slightly orange flushed, with reddish or brownish lines in the tube, narrowly campanulate, rather abruptly narrowed into the base; tube 3–4.3 cm long, the lobes rounded, 1–1.5 cm long, sparsely hairy. Ovary narrowly cylindrical, 3 mm long. Fruit linear, 7–21 × 0.5–1 cm. Seeds 2.2–2.7 cm long; 3–7 mm wide.

var. **stans**;

Leaflets glabrous or with sparse pubescence on midvein beneath and sparsely puberulous on surface beneath. Fig. 6.1–6.4, p. 27.

KENYA. Northern Frontier District: Mathews Range, 22 Dec. 1976, *Ichikawa* 283!; Turkana District: Kacheliba, Dec. 1956, *Wilson* 322!; Nairobi District: Nairobi, near Thika Road House, 10 Dec. 1950, *Verdcourt* 397A!

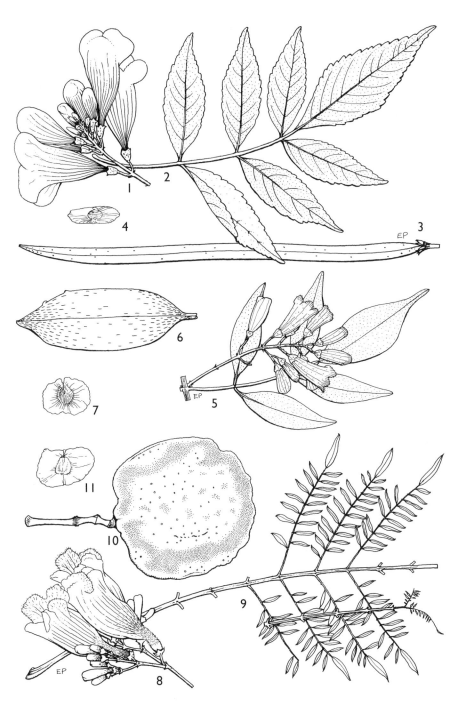

FIG. 6. *TECOMA STANS VAR. STANS* — **1,** habit of shoot with flowers × $\frac{1}{2}$; **2,** leaf × $\frac{1}{2}$; **3,** capsule × $\frac{1}{2}$; **4,** seed × $\frac{1}{2}$. *PANDOREA PANDORANA* — **5,** shoot with leaf and flowers × $\frac{2}{3}$; **6,** capsule × $\frac{2}{3}$; **7,** seed × $\frac{2}{3}$. *JACARANDA MIMOSIFOLIA* — **8,** shoot with flowers × $\frac{2}{3}$; **9,** leaf × $\frac{2}{3}$; **10,** capsule × $\frac{2}{3}$; **11,** seed × $\frac{2}{3}$. 1, 2 from *Perdue* 6339; 3,4 from *Pirozynski* 539; 5 from *Beecher* s.n., Dec. 1962; 6,7, from *Longman* s.n.; 8, 10, 11 from *Ryding* 1157; 9 from *Albers* 61027. Drawn by Emmanuel Papadopoulos.

TANZANIA. Shinyanga District: Old Shinyanga, 25 Mar. 1953, *Welch* 187!; Kigoma District: Kigoma, beach of Lake Tanganyika, 8 Mar. 1964, *Pirozynski* 539!; Dodoma District: Bereko, 28 Mar. 1974, *Richards & Arasululu* 29073!

DISTR. **K** 1, 2, 4; **T** 1, 4, 5; native of America from Florida to Argentina, Bolivia and Peru and West Indies, now cultivated widely and often naturalised

HAB. Roadsides, *Aristida-Harpachne* grassland, rocky places, sandy lake shores; 750–1900 m

SYN. *Bignonia stans* L., Sp. Pl. ed. 2: 871 (1763)

NOTE. *T. stans* (L.) Kunth var. *stans* 'Yellow elder' is one of the most widely grown species of the family and has become naturalised in many parts of the world. Only material is cited which has a fairly clear indication of naturalisation; some of that previously cited as cultivated could be from naturalised populations. It is widely grown in East Africa, e.g. Uganda: Mengo District: Kampala, Makerere University Hill, 5 May 1971, *Lye* 6007!. Kenya: Naivasha District: Lake Naivasha, 26 Nov. 1963, *E. Polhill* 40A; Nairobi District: Closeburn Nurseries, 6 June 1952, *Grahame Bell* 16 & 23 Nov. 1961, *Grahame Bell* H 310/61!. Tanzania: Shinyanga, Nov. 1938, *Koritschoner* 1806; Lushoto District: Amani, 19 Mar. 1973, *Ruffo* 604; Kigoma, 8 March 1964, *Pirozynski* 539; Zanzibar, 1927, *Toms* 30M!.

There are many intermediates with var. *velutina* but typical specimens are distinctive.

var. **velutina** *DC.*, Prodr. 9: 224 (1845); Gentry in Fl. Neotrop. 25(2): 289 (1992). Type: material from Mexico cultivated in Hort. Madrid and Jard. Sarme (G, syn.)

Leaflets ± velvety beneath with longer woolly hairs.

KENYA. Nairobi District: Nairobi, near Thika Road house, 10 Dec. 1950, *Verdcourt* 397B!

DISTR. **K** 4; Mexico

HAB. *Aristida-Harpachne* grassland; ± 1650 m

SYN. *Tecoma mollis* Kunth, Nov. Gen. Sp. 3: 144 (1819). Type: Mexico, Guanajuato, *Humboldt & Bonpland* s.n. (P, holo.)
    Stenolobium molle (Kunth) Seem. in J.B. 1: 91 (1863)

NOTE. *T. stans* (L.) Kunth var. *velutina* DC. is also widely cultivated, e.g. in Uganda: Kigezi District: L. Bunyoni, Habukara 1., 6 May 1963, *Kertland* s.n.. Kenya: Trans-Nzoia District: near Kitale, Kaporetwa, Hort. T.H.E. Jackson, 1 Oct. 1959, *Verdcourt* 2452 A & 2452B; Kiambu District: Ruiru, 6 July 1952, *Kirrika* 206; Nairobi Arboretum, 4 Feb. 1952, *G.R. Williams* 326. Tanzania: Arusha District: Arusha, 16 Dec. 1965, *Leippert* 6173.

NOTE. *T. stans* (L.) Kunth var. *angustata* Rehder. One specimen of this has been seen from Kenya: Nairobi District, Closeburn Nurseries, 3 Mar. 1952, *Grahame Bell* 14. Native of Southern Arizona, part of Texas and Northern Mexico.

Tecoma smithii *Wittmack* (Jex-Blake, Gard. E. Afr. ed. 4: 127 (1957))

Shrub 3–6 m tall with prominently lenticellate shoots. Leaflets 6–8-jugate, opposite, but alternate on some plants, elliptic, 1–3.5(–5) × 0.5–1.5(–2.5) cm; subacute or obtuse. Corolla orange flushed yellow, with bronze markings outside (fide Jex-Blake), narrowly funnel-shaped, 4.5–5 cm long, rather gradually narrowed into the tubular base.

This has been grown in Tanzania (Lushoto District: Amani Nursery, 25 Mar. 1936, *F.M. Rogers* 8 & 5 July 1940, *Greenway* 5959)

NOTE. This was originally named as *Tecoma alata* DC. and *Stenolobium alatum* (DC.) Sprague, a species now considered to be *Tecoma guarume* DC., but Sandwith has determined the above material as *T. smithii* Hort. ex Wittmack a name not mentioned by Gentry. It is probably of hybrid origin.

Tecoma tenuiflora (*DC.*) *Fabris*

Semi-erect shrub 1–2.5 m tall. Leaflets 2–6-jugate, elliptic, 0.5–5 × 0.2–2.5 cm, acute or acuminate, acutely serrulate. Flowers in terminal racemes. Calyx 3–6 mm long with narrow acute teeth. Corolla yellow (in specimens seen but typically red or dark red-orange fide Gentry), very narrowly funnel-shaped, not curved, the tube 4–6 cm long with lobes 4–7 mm long; anthers just exserted from tube. Fruit 6–10(–15) × 0.6–1 cm.

Native of Southern Bolivia and Northwest Argentina. This species has been cultivated in one garden in Kenya and probably elsewhere; Nairobi District: Karen, Hort. Gardner, Feb. 1963, *Gardner* in EAH 12666 & 18 Jan. 1966, *Gillett* 17045 & 1971, *Gillett* 19305.

4. **SPATHODEA**

P.Beauv., Fl. Owar. 1: 46 (1805); Bidgood in Proc. XIIIth Plenary Meeting A.E.T.F.A.T. Malawi 1: 327–331 (1994)

Trees. Leaves opposite or ternate, imparipinnate. Inflorescence a dense terminal raceme with flowers held erect. Calyx spathaceous, boat-shaped, recurved at the apex. Corolla strongly bilaterally symmetrical, very broadly campanulate, narrowing abruptly into a short tube, 2-lipped, upper lip 2-lobed, lower 3-lobed. Stamens 4, didynamous, included to slightly exserted; thecae divaricate. Ovary oblong, Disk annular, lobed. Ovules numerous, multiseriate. Fruit a dry fusiform to compressed capsule; septum flattened. Seeds compressed, heart-shaped in outline, surrounded by a broad membranaceous wing.

A monotypic African genus, but widely cultivated elsewhere.

Spathodea campanulata *P.Beauv.*, Fl. Owar. 1: 47, t. 27 (1805); Seem. in J.B. 3: 332 (1865); Hiern in Cat. Welw. Afr. Pl. 4: 791 (1900); Sprague in F.T.A. 4(2): 529 (1906); U.O.P.Z.: 451, fig. (1949); Heine in F.W.T.A. ed. 2, 2: 386 (1963); Liben in F.A.C., Bignoniaceae: 20 (1977); Hamilton, Uganda Forest Trees: 203 (1981); Gentry in Fl. Cameroon 27: 42 (1984) & in Fl. Gabon 27: 40 (1985); Bidgood in F.Z. 8(3): 62 (1988); Gentry in Fl. Neotropica 25(2): 118, fig. 36 (1992); Diniz in C.F.A. 122. Bignoniac.: 19, t.4 (1993); K.T.S.L.: 592, fig., map (1994); Bidgood in Proc. XIIIth Plenary Meeting A.E.T.F.A.T. Malawi 1: 328, fig. 1 (1994). Type: Nigeria, about 12 km N of Chama, *Palisot de Beauvois* (G, holo)

Tree to 30 m tall, bark grey-brown, variously described as smooth, finely longitudinally fissured or rough; young branches smooth or slightly striate, lenticellate, glabrous to pubescent or tomentose. Leaflets 4–8 pairs, narrowly elliptic to elliptic, (6–)8–15 × 2–7 cm, petiolules if present up to 0.4(–0.7) cm long, base rounded to acute and frequently unequal, margins entire or repand, apex acute to acuminate or cuspidate, upper surface glabrous or with a few scattered hairs mainly on the veins, lower surface glabrous or densely tomentose to tomentose, rarely with only a few scattered hairs, mainly on the veins and showing the dense reticulation, both surfaces with scattered peltate glands, usually with several large glands at the leaf base; terminal leaflet elliptic to broadly elliptic to obovate, base rounded to cuneate and often unequal, apex acute or acuminate (rarely lobed); petiole 8–28(–30) cm long, sometimes with leafy pseudostipules at base, 1–1.5 × 1–1.5 cm. Inflorescence 13–45-flowered; peduncle lenticellate, with prominent scars left by fallen pedicels, glabrous to densely tomentose. Bracts lanceolate, 1–2 × 0.2–4 cm; bracteoles two, at the base of each flower, sometimes elsewhere on the pedicel; bracts and bracteoles with scattered peltate glands ± 1 mm wide. Calyx (3–)3.5–6(–6.3) × (1.2–)2–3.6(–4) cm, tapering, acuminate and recurved at the apex, glabrous or with a few scattered curly hairs, or densely covered with very short velvety hairs or densely tomentose with multicellular hairs. Corolla red to orange tinged with yellow, rarely wholly yellow, glabrous on the outside, sparsely pubescent on the inside, with short crisped glandular and non-glandular multicellular hairs, more towards the base, 7.5–13.5 × (6.2–)7–12 cm; lobes broadly triangular to rounded, 1–3.5 × 2–3.5 cm, margin crinkly; tube 1.2–1.5 × 0.5–0.6 cm, enclosed within calyx. Stamens: anthers

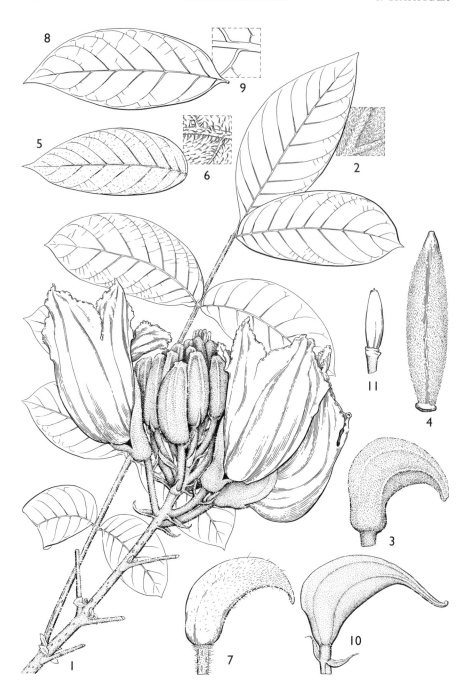

Fig. 7. *SPATHODEA CAMPANULATA SUBSP. NILOTICA* — **1**, inflorescence × ²/₃; **2**, detail of underside of leaflet × 4; **3**, calyx × 1; **4**, young fruit × 1. *SUBSP. CONGOLANA* — **5**, leaflet × ²/₃; **6**, detail of underside of leaflet × 4; **7**, calyx × 1. *SUBSP. CAMPANULATA* — **8**, leaflet × ¹/₂; **9**, detail of underside of leaflet × 4; **10**, calyx × 1; **11**, young fruit × 1. 1–3 from photo; 4 from *Ludanga* 2022; 5, 6 from *Christiansen* 1280; 7 from *Lebrun* 9036; 8–11 from *Latilo & Daramola* 28741. Drawn by Eleanor Catherine. Reproduced with permission from Proc. XIII AETFAT Plenary Meeting.

adnate to the corolla at the mouth of the tube, with the anticous equalling the length of the corolla, the posticous normally shorter; thecae divergent, 6–9 × ± 1 mm, glabrous. Disk thick, lobed, annular. Ovary ± 5 mm long, either glabrous or villous with multicellular hairs; style 5–7 cm long; stigma 2-lobed, flattened, with lobes 5–7 × 2–3 mm, ± elliptic. Fruit held erect, slightly woody, fusiform, 15–23 × 2.5–4.3 cm, compressed, angular (like a flat–bottomed boat when dehisced), lenticellate, glabrous or densely tomentose when young, glabrescent when older; seeds compressed, 0.8–1 × 0.7–1.2 cm; wings hyaline, 1.5–2 × 2–4 cm. Fig. 7, p. 30.

1. Calyx densely hairy, with either short velvety hairs or
   long soft multicellular hairs 2
   Calyx glabrous or with scattered curly multicellular hairs   subsp. **congolana**
2. Leaves densely hairy with long soft multicellular hairs;
   ovary hairy . . . .   subsp. **nilotica**
   Leaves glabrous; ovary glabrous . . . .   subsp. **campanulata**

subsp. **campanulata**; Bidgood in Proc. XIIIth Plenary Meeting AETFAT Malawi 1: 328, fig. 1/G–J (1994)

Leaves glabrous beneath. Calyx densely hairy with short velvety hairs or long soft multicellular hairs. Ovary glabrous. Fig. 7.8–7.12

TANZANIA. Zanzibar: Kusini District, Kibele, 21 Nov. 1999, *Fakih* 463!
DISTR. **Z**, **P**; West Africa, Cameroon, Gabon, Congo-Kinshasa, Angola
HAB. Coral rag forest; 20 m

NOTE. The specimen seen lacks calyx and ovaries but the leaflets are glabrous beneath. Williams in U.O.P.Z: 451 (1949) draws attention to the fact that it is the West African plant which is planted as a roadside tree in Zanzibar and Pemba and specifically mentions the glabrous ovary and leaflets; from Fakih's field note it seems possible it may have become naturalised.

subsp. **nilotica** (*Seem.*) *Bidgood* in Proc. XIIIth Plenary Meeting AETFAT Malawi 1: 330, fig. 1/A–D (1994). Type: Uganda, Bunyoro District: Bunyoro, *Grant* 571 (K!, holo.)

Leaves densely hairy beneath with long soft multicellular hairs. Calyx densely hairy with short velvety hairs or long soft multicellular hairs. Ovary hairy. Fig. 7.1–7.4, p. 30.

UGANDA. Bunyoro District: Masindi, 15 July 1955, *Langdale-Brown* 1459!; Kigezi District: Kinkizi, Kanungu, Mar. 1951, *Purseglove* 3603!; Mengo District: Mulange, Sept. 1920, *Dummer* 4541!
KENYA. Fort Hall District: Thika R., Thika Falls, 18 Feb. 1951, *Greenway & Verdcourt* 8495!; North Kavirondo District: Kakamega Forest Station, 19 Sept. 1949, *Maas Geesteranus* 6305!; Kericho District: about 1.5 km NW of Ngoina Tea Estate Office, 14 Dec. 1967, *Perdue & Kibuwa* 9376!
TANZANIA. Ngara District: Nyakahura to Keza, 18 June 1957, *Gane & MacDonald* 125!, Kigoma District: Mkuti R., July 1955, *Procter* 439! & Mt Livandabe [Lubalisi], 1 June 1997, *Bidgood et al.* 4240!
DISTR. **U** 1–4; **K** 2, 4, 5; **T** 1, 4; Nigeria, Cameroon, Central African Republic, Chad, Congo-Kinshasa, Rwanda, Burundi, Sudan, Ethiopia (fide Vollesen), Angola
HAB. Montane forest and riverine forest, scattered tree grassland with *Combretum*; 750–1500 m

SYN. *Spathodea nilotica* Seem. in J.B. 3: 333 (1865); K. Schum. in P.O.A. C: 363 (1895); Johnston, Uganda Protectorate 1: 68, t. [10] (1902); Sprague in F.T.A. 4(2): 529 (1906); T.T.C.L.: 72 (1949); K.T.S.: 64 (1961)
     S. campanulata sensu I.T.U. ed. 2: 42, t.1 (1952) & Diniz, C.F.A. Bignoniaceae: 21, pro parte quoad *Milne-Redhead* 4051 & K.T.S.L.: 592, fig. (1994), *non* P.Beauv. sensu stricto

NOTE. There is an unlocalised *Hannington* sheet at K; the nearest he got to the known indigenous area of this plant was Tabora where it has presumably never been wild and must have been cultivated in 1883–4.
     Known as the Nandi Flame Tree, Uganda Flame Tree or Tulip Tree this is widely cultivated in East Africa. Uganda (Kampala, July 1963, *Kertland* s.n.). Kenya (Nairobi Arboretum, 23 June 1952, *G. R. Williams* 467 & Nairobi, Hort. Stewart, July 1963, *J. Stewart* 767). Tanzania

(Lushoto District: Lushoto, near Police Station, 6 Apr. 1967, *Semsei* 4220; Ufipa District: Sumbawanga, Upendo View Country Club, 26 Nov. 1994, *Goyder et al.* 3839; Mbeya District: Mbeya, Mt. Livingstone Hotel, 29 Nov. 1994, *Goyder et al.* 3842; Lindi District, Rondo Plateau, St. Cyprian's College, 20 Feb. 1991, *Bidgood et al.* 1681). In areas where it occurs wild it is sometimes difficult to decide whether it is cultivated or not. A form with rich buttercup yellow flowers is mentioned in I.T.U.: 42 (1952) as occurring in Uganda (Bugishu and Mengo) and Jex-Blake (Gard. E. Afr., ed. 4: 331 (1957)) mentions such a tree as having been found near Kampala and propagated by root cuttings. It has been grown in Kenya (Machakos District: Kayata Estate, Hort. Delap. 15 Jan. 1963, *Bally* 12665; Nairobi District; Fairview Hotel, 7 Nov. 1994, *Goyder et al* 3712).

subsp. **congolana** *Bidgood* in Proc. XIIIth Plenary Meeting AETFAT Malawi 1: 330, fig. 1E, F (1994). Type: Congo-Kinshasa, Rutshuru, *Lebrun* 9036 (BR!, holo; K!, iso.)

Leaves with less dense multicellular hairs. Calyx glabrous or with scattered curly multicellular hairs. Fig. 7.5–7.7, p. 30.

Tanzania. Kigoma District: Gombe Stream National Park, Mitumba Camp, 8 Feb. 2000, *Gobbo* 619!
Distr. **T** 4; Congo-Kinshasa, Burundi, Rwanda
Hab. Disturbed grassland and forest margins; ± 780 m

Note. Another specimen is cited by Bidgood (op. cit.) from the same place (Gombe Stream National Park, 2 Apr. 1970, *Clutton Brock* 568!) and annotated as an intermediate between subsp. *campanulata* and subsp. *congolana.*

5. **MARKHAMIA**

Baill., Hist. Pl. 10: 47 (1888)

Shrubs or trees. Leaves imparipinnate, with well developed foliaceous or subulate pseudostipules. Inflorescences terminal panicles or racemes. Calyx spathaceous, cuspidate at the apex. Corolla campanulate or campanulate-funnel-shaped, tubular at the base, 5-lobed, 2-lipped. Stamens, 4, didynamous, included; anther thecae connate above, diverging towards the base. Disk annular or cupular. Ovary oblong with 2 locules; ovules numerous. Capsule linear or curved, flattened, 2-valved, splitting loculicidally, glabrous to velvety (or in an Asian species floccose-woolly). Seeds narrowly oblong in outline, bilaterally winged.

A genus of six species, four African, the others in China, Indochina and Burma. Sprague (F.T.A. 4(2): 522–528 (1906) & K.B. 1919: 302–314 (1919)) recognised about a dozen species but a vast amount of additional material has shown that many of them cannot be maintained.

1. Leaflets densely velvety-woolly beneath; pseudostipules
   subulate; flowers bright yellow; fruits densely velvety
   tomentose with golden hairs . . . .   3. *M. obtusifolia*
   Leaflets glabrous to velvety beneath; pseudostipules leafy;
   flowers golden yellow or pale yellow and reddish purple;
   fruits glabrous to pubescent or if velvety then hairs not
   golden 2
2. Leaves glabrous or pubescent to velvety; flowers paler yellow
   and reddish purple; mostly coastal species (0–1350 m) . .   1. *M. zanzibarica*
   Leaves glabrous save for domatia; flowers golden yellow;
   upland species (1500–1900 m) . . . .   2. *M. lutea*

1. **Markhamia zanzibarica** (*DC.*) *Engl.,* Abh. Königl. Preuss. Akad. Wiss. Berl. 1894: 16 (1894); K. Schum. in E. & P., Pf. 4 (3b): 242 (1895) & in P.O.A. C: 363 (1895); Sprague in F.T.A. 4 (2): 523 (1906) & in K.B. 1919: 313 (1919); T.T.C.L.: 72 (1949);

K.T.S.: 64 (1961); Haerdi in Acta Trop., suppl. 8: 149 (1964); Liben, F.A.C. Bignon.: 26 (1977); Vollesen in Opera Bot. 59: 78 (1980); Diniz in F.Z. 8(3): 74, t. 14A (1988) & C.F.A. 122, Bignon.: 21 (1993); K.T.S.L.: 592, map (1994); Tardelli & Settesoldi in Fl. Som. 3: 304, fig. 210 (2006). Type: Zanzibar, *Bojer* s.n.(G, holo.)

Shrub or much branched small tree, 2.4–9(?–15) m tall, readily regenerating after cutting; bark pale grey, smooth or rough, peeling; young branches usually lenticellate, minutely scaly. Leaves up to 35 cm long; leaflets 1–4-jugate or rarely reduced to the terminal leaflet, elliptic, elliptic-lanceolate, obovate, ovate or almost round, 2–25(–33) × 2–13 cm, acute to long-acuminate or obtuse to rounded at the apex, cuneate to rounded at the base, almost entire to toothed or rarely almost lobed, glabrous in typical coastal specimens save for pubescent axillary domatia usually present beneath, pubescent to densely velvety in (mostly) inland material, densely to sparsely lepidote; petiole 2–7(–9) cm long, sometimes together with rhachis faintly winged; petiolules 0–5 mm long; pseudostipules if present rounded to reniform, 0.5–2.5 cm long and wide. Flowers in terminal or axillary rather narrow lepidote panicles 2–11(–30) cm long; secondary peduncles up to 1 cm long, pedicels 0.5–1.5(–2) cm long with bracts 2–5(–7) mm long. Calyx 1–1.5 cm long, rounded, cuspidate or uncinate at the apex, split to about 8 mm from base, usually glabrous but densely hairy in some variants. Corolla yellow-green to yellow, the tube ± densely speckled with liver red or reddish purple spots and blotches and the lobes dull brown to reddish purple inside and speckled outside (often mostly purplish or brown with reduced yellow markings); tube campanulate to funnel-shaped, (1.8–)2–3(–4) cm long, limb ± 2–lipped, 5-lobed, the lobes rounded, 1–1.5 cm long and wide; discoid glands present at top of tube and on the lobes, probably greenish in life but dark when dry. Capsule straight or curved, 14–72 × 0.9–1.8 cm, glabrous, usually with very conspicuous whitish lenticels; seeds 2–4 cm long, 4–6 mm wide including wing.

KENYA. Kwale District: Mrima Hill, 5 Mar. 1977, *R.B. & A.J. Faden* 77/673!; Kilifi District: 6.5 km N of Malindi, Sabaki, 3 Nov. 1961, *Polhill & Paulo* 712!; Lamu District: 4 km S of Shakani ruins, NE of Kiunga, 5 Apr. 1980, *Gilbert & Kuchar* 5876!
TANZANIA. Pangani District: Bushiri, 18 Sept. 1950, *Faulkner* 704!; Morogoro District: Morogoro, 14 Dec. 1933, *B.D. Burtt* 4998!; Masasi District; just W of Bangala R., 16 Dec. 1955, *Milne-Redhead & Taylor* 7677!; Zanzibar: Kokotoni, Oct. 1873, *Hildebrandt* 977!
DISTR. **K** 7; **T** 1–8; **Z**, **P**; Somalia, Angola, Zambia, Malawi, Mozambique, Zimbabwe, Botswana and northern South Africa; cultivated in northern Australia
HAB. Thorn-bush, coastal thicket and woodland on coral rag, *Adansonia–Dobera* woodland, deciduous bushland and forest edges; 0–1350 m

SYN. *Spathodea zanzibarica* DC., Prodr. 9: 208 (1845); Klotzsch in Peters, Reise Mossamb., Bot. 1: 191 (1861)
   S. acuminata Klotzsch in Peters, Reise Mosamb., Bot. 1: 191 (1861). Type: Mozambique, Tete, Rios de Sena, *Peters* s.n. (B†, holo.)
   S. puberula Klotzsch in Peters, Reise Mossamb., Bot. 1: 192 (1861). Type: Mozambique, Tete, Rios de Sena, *Peters* s.n. (B†, holo.)
   Muenteria stenocarpa Seem. in J.B. 8: 329, t.36 (1865). Types: Angola, Pungo Andongo, Mata de Cabondo, *Welwitsch* 483 (LISU, syn; BM!, COI, K!, isosyn.) & Golungo Alto, Luinha R. and near Cambondo, *Welwitsch* 482 (LISU, syn.; BM!, K!, isosyn.)
   Dolichandrone hirsuta Bak. in K.B. 1894: 31 (1894). Type: Mozambique, Tete, banks of the lower Zambesi, *Kirk* s.n. (K!, holo.)
   D. latifolia Bak. in K.B. 1894: 31 (1894). Type: Kenya, Nyika Country, *Wakefield* s.n. (K!, holo & iso.)
   D. stenocarpa (Seem.) Bak. in K.B. 1894: 31 (1894)
   Markhamia acuminata (Klotzsch) K.Schum. in P.O.A. C: 363 (1895); Sprague in F.T.A. 4(2): 524 (1906) & in K.B. 1919: 313 (1919); T.T.C.L.: 71 (1949); Pardy in Rhod. Agric. J. 53: 58, illustr. (1956); F.F.N.R.: 379 (1962); Palmer & Pitman, Trees S. Afr. 3: 2009, figs. (1967); Vollesen in Opera Bot. 59: 77 (1980); Palgrave, Trees S. Afr.: 831 (1981)
   M. puberula (Klotzsch) K.Schum. in E. & P., Pf. 4(3b): 242 (1895); Sprague in F.T.A. 4(2): 523 (1906) & in K.B. 1919: 312 (1919)

M. stenocarpa (Seem.) K.Schum. in E. & P., Pf. 4(3b): 242 (1895); Sprague in F.T.A. 4 (2): 524 (1906) & in K.B. 1919: 313 (1919)

M. infundibuliformis K.Schum. in E. & P., Pf. 4(3b): 242 (1895) & in P.O.A. C: 363 (1895); Sprague in F.T.A. 4(2): 524 adnot. (1906). Type: Tanzania, Singida District, Ussure, *Fischer* 460 (B†, holo.)

NOTE. As treated here, following Diniz in F.Z., this species is extremely variable particularly in the shape, toothing and indumentum of the leaves. However, it has not been possible to divide it satisfactorily into infraspecific entities. Despite Diniz's comment about the leaves ("or even glabrous") the species is typically glabrous, material from the type locality and lowland coastal areas of Kenya and Tanzania having almost uniformly glabrous somewhat thick more evergreen leaves, with rare exceptions (e.g. *Tanner* 9, Pangani District, Msubugwe Forest and *Harris* 1544, Uzaramo District, 20 km NW of Dar es Salaam, Wazo Hill). Away from the coast material has puberulous to velvety leaves, glabrous to velvety calyces and much variation in leaf-size but no geographical correlation can be found. Very velvety variants could be given a varietal name based on *Dolichandrone hirsuta* Bak. A distinctive variant is *Greenway & Kanuri* 14441 (Iringa District, Ruaha National Park, Kinyantupa track) with rounded or oblanceolate leaflets 10–12 cm wide and distinctly lobed-crenate; other material from the same area is similar. A detailed field study of the species throughout its range may reveal characters not evident from herbarium material. *Richards* 24747 (**T** 2) reports flowers pale cream. Jex-Blake, Gard. E. Afr.: 350 (1957) mentions this, but we have not been able to trace any cultivated material.

2. **Markhamia lutea** (*Benth.*) *K.Schum.* in E. & P., Pf. 4 (3b): 242 (1895); Sprague in Hook. Ic. 28: sub t. 2800 (1905) & in F.T.A. 4(2): 525 (1906) & in K.B. 1919: 311 (1919); F.P.NA. 2: 248 (1947); Heine in F.W.T.A. ed. 2, 2: 387 (1963); Liben, F.A.C. Bignon.: 28, fig. 2A (1977); Gentry in Fl. Cameroun 27: 36, t. 10 (1984) & in Fl. Gabon 27: 34, t. 7 (1985); Blundell, Wild Fl. E. Afr.: 381, t. 396 (1987); K.T.S.L: 592, fig., map (1994). Type: Fernando Po, *Vogel* 60 (K!, lecto.)

Tree (5–)15–21(–24) m tall or sometimes shrubby 1.2–4.5 m tall, regenerating and coppicing easily; trunk sometimes divided from base, sometimes fluted; bark grey-black or reddish brown, smooth or rough, ridged; branches usually densely lenticellate; wood reddish. Leaves up to 35 cm long; leaflets (2–)3–6-jugate, elliptic to obovate, 4.5–21 × 4–9 cm, cuneate to rounded at the base, acuminate at the apex, glabrous but densely lepidote above, more sparsely so beneath, often with pubescent axillary domatia beneath; petiole 6–12 cm long; petiolules up to 5(–10) mm long or ± absent; pseudostipules ± round, 2–3 cm diameter. Flowers scented, in often narrow terminal panicles up to 20 cm long and wide; pedicels ± 5(–10) mm long; bracts triangular, 1–5 mm long, 1 mm wide. Calyx spathaceous, 1.8–2.6(–3) cm long, 0.7–1.4 cm wide, woolly when young, becoming glabrous, lepidote and with glands on the margin opposite the split, often very distinctly uncinate at the tip. Corolla golden yellow with brownish purple or red veins or spots at the throat; tube (2–)3–4.5 cm long, up to 2.5 cm wide at the throat; limb 2-lipped, 5-lobed, the lobes (1–)1.5–2.5 cm long and wide, with distinct glands but varying in number. Capsule curved, 35–80(?–105) cm long, 1–2 cm wide, compressed, glabrous but lepidote; seeds 2.5–3.5 cm long, 6–8 mm wide including the wing.

UGANDA. W Nile District: S Panyimur, 25 May 1954, *Alonzie* 14!; Kigezi District: Kinkizi, Kanungu, Mar. 1951, *Purseglove* 3604!; Mengo District: Bugerere, Bale, 28 June 1956, *Langdale-Brown* 2119!

KENYA. Nairobi District: Nairobi, Karura Forest, 12 Nov. 1976, *Kahuranang'a & Mungai* 131!; Embu District: Emberre, 7 Oct 1933, *M.D. Graham* 2271!; South Kavirondo District: near Kisii, 31 Jan. 1964, *Brunt* 1440!

TANZANIA. Mwanza District: Geita, Uzinza, 7 June 1937, *B.D. Burtt* 6588!; Lushoto District: E Usambaras, Amani, 2 Oct. 1955, *Tanner* 2273!; Mpanda District: 104 km S of Kigoma, Mugombasi, 31 Aug. 1959, *Harley* 9473!

DISTR. **U** 1–4; **K** 1, 3–5; **T** 1, 3, 4 (planted in **T** 2); West Africa from Ghana to Cameroon, Congo-Kinshasa, Rwanda, Burundi

HAB. Wooded grassland, submontane and riverine forest, evergreen rain forest; 1500–1900 m

SYN. *Spathodea lutea* Benth. in Nig. Fl.: 461 (1844) pro parte
    Dolichandrone platycalyx Bak. in K.B. 1894: 30 (1894). Type: Uganda, without precise
        locality, *Wilson* 119 (K!, holo.)
    D. hildebrandtii Bak. in K.B. 1894: 31 (1894). Type: Kenya, Kitui, *Hildebrandt* 2732 (K!,
        holo.)
    Markhamia platycalyx (Bak.) Sprague in Hook. Ic. 28: t. 2800, fig. 1–7 (1905) & in F.T.A. 4
        (2): 525 (1906) & in K.B. 1919: 311 (1919); T.T.C.L.: 72 (1949); I.T.U. ed. 2: 41, fig. 8,
        photo 5 (1952); Jex-Blake, Gard. E. Afr.: 222, 335 (1957); Wimbush, Cat. Kenya Timbers:
        55 (1957); K.T.S.: 62, fig. 13 (1961); Hamilton, Uganda Forest Trees: 202 (1981)
    M. hildebrandtii (Bak.) Sprague in Hook. Ic. 28: t. 2800, fig. 9 (1905) & in F.T.A. 4 (2): 526
        (1906) & in K.B. 1919: 311 (1919); T.T.C.L.: 71 (1949); Jex-Blake, Gard. E. Afr..: 222
        (1957); Wimbush, Cat. Kenya Timbers: 54 (1957); K.T.S.: 62 (1961)

NOTE. In a herbarium memorandum and various annotations on herbarium covers J. Gillett
sank *M. hildebrandtii* and *M. platyclayx* together and later accepted Liben's published
statement that the latter was conspecific with *M. lutea*. Gentry published this synonymy in
the references given above. We have followed this but with some hesitation. Material from
eastern Kenya undoubtedly differs to some degree from material from western Kenya and
Uganda. *M. hildebrandtii* was distinguished by Sprague as having a distinctly uncinate calyx,
more infundibuliform flowers and very few large floral glands. Sandwith in the mid-1900's
kept all three separate. Typical *M. platycalyx* with wider almost clavate calyx much larger
more campanulate corolla with often numerous floral glands appears very different. In West
Africa, however, uncinate or at least narrowly acuminate calyces reappear. A specimen
Mwangangi 1945/B (Kenya, Machakos District, near Nunguni, Kilungu Forest), which on
geographical grounds would be expected to be nearest *M. hildebrandtii* actually has the calyx
of *M. platycalyx*. Being a well known timber it is possible this was cultivated from non-local
seed. The species would repay study in the field throughout its range to see if subspecies
could be recognised. A specimen *Kayombo* 1659 (**T** 6, Ulanga District, Nambiga Forest
Reserve, 5 km W of main road from Iragua to Itete, 380 m) differs from this species in the
leaflets having much more widely spaced lateral nerves (± 6 instead of 8–14). It may
represent a distinct taxon and is from a low altitude; fruits and a larger range of leaves
showing variation are needed. The large yellow corolla 6 cm long and calyx 2 cm long agree
with *M. lutea*.
    This species is extensively planted particularly in Uganda as it is commonly used as a local
timber. Also grown as an ornamental in Kenya: Nairobi Arboretum, 14 July 1952, *G.R.
Williams* 483 & National Museum grounds, 11 Sept. 1952, *G.R. Williams* 521; and Tanzania:
Arusha District: Narok Camp, 22 March 1951, *J.F. Hughes* H58; Lushoto District: Amani,
Kiumba Plantation, 21 Oct. 1930, *Greenway* 2546.

3. **Markhamia obtusifolia** (*Bak.*) *Sprague* in K.B. 1919: 312 (1919); T.T.C.L.: 71
(1949); K.T.S.: 62 (1961); F.F.N.R.: 379 (1962); Liben. F.A.C., Bignon: 32, t.8, A–C
(1977); Vollesen in Opera Bot. 59: 78 (1980); Palgrave, Trees S. Afr.: 831 (1981);
Diniz in F.Z. 8(3): 76, t. 143 (1988) & C.F.A. 122 Bignon.: 23 (1993). Type:
Mozambique, Shupanga, *Kirk* s.n. (K!, lecto.), chosen by Diniz

Shrub 1.5–3.5 m tall or small tree 5–15(?–20) m tall*, freely regenerating; bark
pale brown or grey, smooth or somewhat striated or ridged in larger trees; young
branches golden brown velvety tomentose, soon glabrescent. Leaves 18–56 cm
long; leaflets (1–)3–5(–6)-jugate, oblong, elliptic, ovate or obovate,
(4.5–)6.5–18(–23) × 2.5–9.5(–13.5) cm, cuneate, rounded or subcordate at the
base, entire or very slightly serrate, rounded or obtuse to very shortly acuminate at
the apex, puberulous or pubescent above, densely woolly velvety beneath; petiole
2–9.5 cm long; petiolules 0–2 mm long; pseudostipules subulate, 0.5–1.3 cm long,
tomentose. Flowers in dense tomentose terminal panicles, at first tight clusters ±
8 cm long, later 10–30(–40) cm long; pedicels 1–2.8 cm long, bracts 0.5–1(–1.5) cm
long; buds ± uncinate; calyx 1.9–3.3(–4) cm long, 1.5–2.5 cm wide, obtusely
cuspidate or uncinate, split to about 4–8 mm from the base, densely golden
tomentose. Corolla bright yellow or orange-yellow with maroon guide lines in

---

* Schlieben gives 10–20 m; a field note mentioning 45 m is clearly an error.

FIG. 8. *MARKHAMIA OBTUSIFOLIA* — **1,** habit of shoot × ²/₃; **2,** flower × ²/₃; **3,** fruit × ²/₃; **4,** seeds × ²/₃. 1 from *Goyder et al.* 3795; 2 from *Geilinger* 3234; 4 from *Boaler* 241. Drawn by Emmanuel Papadopoulos.

throat and maroon stripes or spots on 3 lower lobes; tube 3.5–4.5 cm long with lobes 1.5–3 cm long and wide and upper part of tube with conspicuous glands. Capsule nearly straight or curved, (20–)39–87 cm long, 1.4–2.8 cm wide, with prominent longitudinal median ridge on valves and marginal ones near sutures, densely golden velvety tomentose; seeds narrowly oblong, 3–4 cm long, 0.7–1.2 cm wide including the wing. Fig. 8, p. 36.

TANZANIA. Lushoto District: Korogwe, 19 Dec. 1962, *Archbold* 85!; Morogoro District: ± 29 km N of Morogoro along Dodoma road, 14 Mar. 1988, *Bidgood et al* 428!; Lindi District: Nachingwea, 17 Nov. 1949, *Anderson* 546!

DISTR.* **T** 1–8; Congo-Kinshasa, Rwanda, Burundi, Angola, Zambia, Malawi, Mozambique, Zimbabwe, Botswana, Namibia, South Africa; cultivated in Nairobi

HAB. Wooded grassland, bushland, *Brachystegia* woodland, *Commiphora* thicket, lakeshore forest, often on rocky hills; (0–)250–2050 m

SYN. *Dolichandrone obtusifolia* Bak. in K.B. 1894: 31 (1873)**
   Markhamia lanata K.Schum. in E. & P., Pf. 4 (3b): 242 (1895); Sprague in Hook. Ic. 28: t. 2800, fig. 8 (1905); Sprague in F.T.A. 4(2): 527 (1906). Type: Tanzania, unspecified syntypes (B†) (see note)
   M. paucifoliolata De Wild. in Ann. Mus. Congo, Bot., sér. 4(1): 131 (1903). Type: Congo-Kinshasa, Lukafu, *Verdick* 54 (BR, holo.)
   M. verdickii De Wild. in Ann. Mus. Congo, Bot., sér. 4(1): 132 (1903). Type: Congo-Kinshasa, Lofoi, *Verdick* s.n. (BR, holo.)
   M. tomentosa sensu K.Schum. in Engl., Abhandl. K. Preuss. Akad. Wiss. Berlin, Phys. Math. K 1894: 34, 39 (1894) pro parte & in E.J. 28: 480 (1900), *non* (Benth.) K.Schum.

NOTE. K. Schumann merely gives 'aus den Steppen Ostafrikas' for his *M. lanata* but it is almost certain that *Fischer* 461 (Kagehi), *Buchwald* 697 (Usambaras), *Holst* 2228 (E Usambaras, Magila) and *Busse* 50 (probably **T** 3) are all syntypes; duplicates of the last three at Kew are isosyntypes. *M. obtusifolia* has been cultivated as an ornamental in Kenya (Nairobi, 1942, *Bally* 6168).

6. **STEREOSPERMUM**

Cham. in Linnaea 7: 720 (1832)

Erect trees or shrubs. Leaves imparipinnate; leaflets with main lateral veins opposite, sub-opposite or alternate, domatia absent. Inflorescence a terminal lax panicle. Calyx campanulate or tubular, variously lobed. Corolla slightly bilaterally symmetrical, infundibuliform, 2-lipped, the upper lip 2-lobed, the lower lip 3-lobed. Stamens 4, didynamous, included; thecae divergent; staminode present. Disk cupular to annular. Ovary of 2 locules with many ovules; stigma 2-lobed, flattened. Fruit long, terete, often twisted; septum thick, cavity with alternate notches accommodating the ellipsoid seeds, the membranes of which are flush to the septum surface.

A genus of 15–20 species in tropical Asia, Africa and Madagascar.

Stereospermum kunthianum *Cham.* in Linnaea 7: 721 (1832); K. Schum. in E. & P., Pf. 4(3b): 242 (1895) & in P.O.A. C: 364 (1895); Sprague in F.T.A. 4(2): 518 (1906); T.T.C.L.: 73 (1949); I.T.U ed. 2: 42, col. pl. 2, fig. 9 (1952); K.T.S.: 66 (1961); F.F.N.R.: 379 (1962); Heine in F.W.T.A. ed. 2, 2: 386 (1963); Liben in F.A.C., Bignon.: 14 (1977); Palgrave, Trees S. Afr.: 832 (1981); Diniz in F.Z. 8(3): 77, t. 15 (1988); K.T.S.L.: 593, fig., map (1994). Type: Senegal, ex Herb. Kunth., possibly *Perrottet* 499 (?LE, holo.; G!, iso.)

  * KTS records the species from Nyanza and Diniz in FZ gives Kenya in her distribution. We have seen no wild material from Kenya. *J. Bally* 6168 in CM 17981 was grown in Nairobi from seed from Mwanza (**T** 1).
  ** One syntype *Kirk* s.n. from Bagamoyo is the only specimen reported from near sea level; it could quite possibly have come from further inland.

Erect shrub or small tree 2–20 m tall, glabrous to tomentose, overall indumentum without glands, bark grey to whitish, rarely dark brown, smooth and flaking in plaques; young branches smooth or striate, frequently lenticellate, glabrous to tomentellous. Leaves (7–)9–43 cm long including petiole (0.7–)1.5–9(–10) cm, rachis slightly striate or grooved, frequently lenticellate, hairs not glandular; leaflets rotund to elliptic or narrowly elliptic, occasionally obovate, (2.5–)3–17.5 × (1.2–)2–5.7(–10.5) cm; terminal leaflet usually larger, apex retuse to cuspidate, sometimes unequal, base rounded to cuneate, frequently unequal, margins entire or crenulate-sinuate, occasionally crenate or serrate, hairs occasionally confined to veins and midrib, secondary venation reticulate, lower surface usually with fine gland dots throughout and large scattered glands either side of the midrib towards the base. Inflorescence up to 48 cm long including peduncle, most parts including calyx frequently with long white hairs when young, glabrous or pubescent to tomentose, sometimes lenticellate, pedicels frequently with small glands just below the calyx. Flowers fragrant, often precocious, bracts 3–4 × 0.5–0.7 mm, corolla falling off soon. Calyx (4–)5–10(–12) mm long, campanulate, truncate with sinuate margins, or lobes broadly triangular to obtuse, up to 3 mm long, frequently with small mealy glands, usually with larger scattered peltate glands towards the apex, occasionally viscid. Corolla pinkish-white, reddish-pink, pale lilac to purple, usually with darker lines into throat, (2.3–)3–6(–6.5) cm long, lobes ± ¹/₃ the length of the tube, broadly ovate, margins undulate, densely pubescent outside and on the palate, with glandular hairs in the lower part of the tube inside. Stamens: filaments of anticous stamens 1–1.5 cm long and of posticous 0.7–1.2 cm long, glabrous, glandular-pubescent at base; thecae ± 2 mm long. Pistil with cupular disk at base, slightly lobed. Fruit 15–60 cm long, glabrous or puberulous, finely lenticellate, often with scattered peltate glands. Seeds 5 × 3 × 3 mm, wings 6–7 × 2–3 mm. Fig. 9, p. 39.

UGANDA. Lango District: near Aloro, Jan. 1931, *Brasnett* F.D. 53!; Toro District: N Ruwenzori Forest Reserve on Fort Portal to Semliki valley road, before Buranga Falls, 16 Jan. 1937, *D. Wood* 839!; Busoga District: Bugabula area, about 5 km S of Mulima on Lake Kyoga, Jan. 1931, *Brasnett* F.D. 54!

KENYA. Turkana District: about 8 km from Moroto turn-off on road to Lodwar, 9 Oct. 1952, *Verdcourt* 801!; Trans-Nzoia District: Kitale to Kapenguria, 3 Feb. 1947, *Bogdan* 265!; Meru District: track from Kieiga Forest to Ruiri School, S of Thuuri Forest, 25 June 1974, *R.B. & A.J. Faden* 74/879!

TANZANIA. Shinyanga District: Mantini Hills, Ngaramhuri Rocks, Sept. 1935, *B.D. Burtt* 5301!; Pangani District: Bushiri, 25 Nov. 1950, (fl) & 20 Jan. 1951 (fr.) *Faulkner* 701!; Lindi District: Lindi–Mtwara road, near Ruhokwe Village, 24 Oct. 1978, *Magogo & Rose Innes* 414!

DISTR. **U** 1–4; **K** 2–5, 7; **T** 1–8; widespread in tropical Africa

HAB. Grassland with scattered trees, *Brachystegia* and other woodland and bushland; 20–2050 m

SYN. *Bignonia lanata* Fresen. in Flora 21: 607 (1838). Type: Ethiopia, *Salt* s.n. (BM, syn.); *Ruppell* s.n. (FR, syn.)

    B. discolor R.Br. in Salt, Append.: 94 (1814) *nomen nudum*, based on a *Salt* specimen from Ethiopia

    Stereospermum dentatum A.Rich., Tent. Fl. Abys. 2: 58 (1851); K. Schum. in E.& P. Pf. 4 (3b): 242, fig. 92/H–K (1895) & in P.O.A. C: 364 (1895). Type: Ethiopia, Adua, Mt Scholoda, *Schimper* I.308 (P, syn.; K!, isosyn.) & Selleuda, *Schimper* II. 808 (P, syn.)

    S. integrifolium A.Rich., Tent. Fl. Abys. 2 59 (1851); K. Schum. in E. & P. Pf. 4 (3b): 242 (1895) & in P.O.A.C: 364 (1895). Type: Ethiopia, Shire, Aderbati, *Quartin Dillon & Petit* s.n. (P, holo.)

    Dolichandrone smithii Bak. in K.B. 1894: 30 (1894). Type: Tanzania, Kilimanjaro, *C.S. Smith* s.n. (K!, holo.)

    Stereospermum molle K.Schum. in E.& P. Pf. 4 (3b): 242 (1895). Type: Sudan, Mittu lands on White Nile, collector not stated but must be *Schweinfurth* (B, syn.) (several numbers at K could be syntypes)

    S. discolor K.Schum. in E. & P. Pf. 4 (3b): 242 (1895), *nomen* (based on *Bignonia discolor* R. Br.).

    S. kunthianum Cham. var. *dentatum* (A.Rich.) Fiori, Bochie Piantae Legnosae dell' Eritrea: 341 (1912)

NOTE. *Holst* 2229 (K!) (Tanzania, Usambara foothills, Magila is labelled *S. kunthianum* var. *fulva* K.Schum. but this name does not appear to have been published. Steudel, Nomencl. ed. 2 1: 205 (1840) lists *Bignonia lanata* R.Br. in Salt, Append. but it is not mentioned there.

Fig. 9. *STEREOSPERMUM KUNTHIANUM* — **1,** inflorescence × ²/₃; **2,** leaf with capsule × ²/₃; **3,** flower opened × ²/₃; **4,** seeds × ²/₃. All from *Friis et al.* 9159. Drawn by Emmanuel Papadopoulos.

7. **FERNANDOA***

Seem. in J.B. 3: 330 (1865) (as '*Ferdinandia*') & in J.B. 4: 123 (1866) & J.B. 8: 280 (as '*Ferdinandoa*') & in J.B. 9: 81 (1871); Milne-Redh. in K.B. 4: 170 (1948); Bidgood & Brummitt in Taxon 42: 675–678 (1993); Bidgood in K.B. 49: 381–390 (1994)

Trees or shrubs. Leaves imparipinnate, frequently undeveloped at time of flowering. Flowers single or in clusters on old branches (cauliflorous), or in axillary clusters or racemes. Calyx campanulate, irregularly 3–5-lobed. Corolla large, showy, brownish red to orange-red or yellow, tube campanulate (more broadly so in the African than in the Madagascan species), narrowing abruptly into the calyx, limb slightly bilabiate. Stamens slightly exserted, thecae divaricate; staminode present. Ovary narrowly cylindrical, bilocular. Disk cupular. Capsule long, subterete, dehiscing at right angles to the flattened septum. Seeds numerous, compressed, with lateral hyaline wings.

A genus of eight species, five from Africa and three from Madagascar. *Haplophragma* (a genus of six species from SE Asia and Sumatra) was united with *Fernandoa* by van Steenis but the two appear adequately distinct.

1. Flowers in racemes with thick peduncles 4–21 cm long; thecae
    9–12 mm long (corolla scarlet or orange-red) . . . .   2. *F. magnifica*
    Flowers single or in clusters on old branches or in axillary
    clusters or racemes with slender peduncles up to 2.2 cm
    long; thecae 5–8 mm long 2
2. Corolla yellow; median leaflets less than twice as long as broad;
    leaflets uniformly hairy beneath . . . .   1. *F. lutea*
    Corolla orange-red or red to reddish brown; median leaflets
    more than twice as long as broad; leaflets glabrous or only
    hairy on main veins beneath . . . .   3. *F. abbreviata*

1. **Fernandoa lutea** (*Verdc.*) Bidgood in K.B. 49: 383 (1994). Type: Tanzania, Rondo Plateau, *Bryce* in *Eggeling* 6044 (EA!, holo.; K!, iso.)

Tree 20–30 m tall, bark grey-brown, striate or rough, slash red-brown, white underneath (fide Eggeling); young branches grey-brown, striate, puberulous, with scattered small peltate glands, denser towards the apex, older branches grey-brown, deeply striate with a gnarled appearance, frequently lenticellate. Leaves 3–5-jugate, ± 24 × 18 cm, including petiole of 3–5 cm; rachis striate, puberulous, glandular when young; leaflets ovate, the median less than twice as long as broad, base subcordate and usually unequal, apex acuminate; lateral leaflets 6.5 × 3–6 cm including petiolule 1–3 mm; terminal leaflet 6–8.3 × 3.5–6 cm, young leaflets minutely glandular on both surfaces, mature leaflets glabrous above apart from the sometimes puberulous midrib, uniformly pubescent beneath, both surfaces with a few large scattered glands. Flowers precocious, borne singly or in 2–3-flowered inflorescences with peduncles of 1–7 mm long, either from the axils of the old leaves or cauliflorous; pedicels ± 3–3.5 cm long, striate, puberulous, sometimes glandular just below the calyx. Calyx broadly campanulate, 2–3 × 2.5–3.5 cm, 3–4-lobed, chartaceous, minutely puberulous, with scattered large glands and sometimes small peltate glands toward the base; lobes elliptic to broadly elliptic, obtuse to rounded, apiculate, 1–2 × 1–2 cm, margins scarious. Corolla yellow with purple veins, very broadly campanulate, 6–11 × 7.5–12.5 cm, 5-lobed, the lobes broadly ovate to cordiform, overlapping, margins slightly undulate, tube 3–4 cm long; filaments 4–6.5 cm long,

* The troubled history of the generic name and its numerous variants are explained by Bidgood and Brummitt in the reference cited.

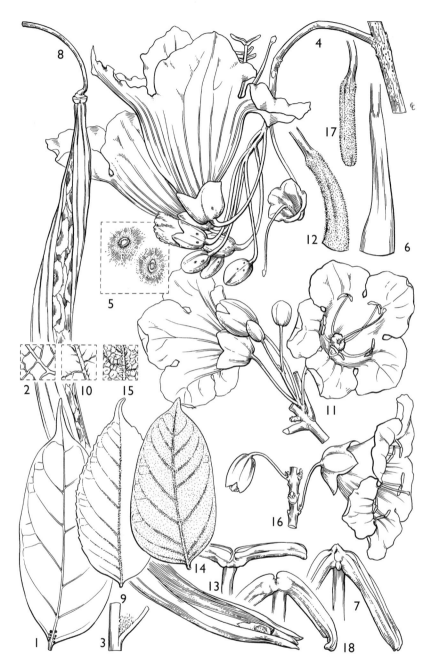

FIG. 10. *FERNANDOA MAGNIFICA* — **1**, leaflet × ¹/₂; **2,** detail of indumentum on underside of leaf × 10; **3**, domatium × 6; **4**, inflorescence × ¹/₂; **5**, calyx glands × 10; **6**, ovary × 3; **7**, anther × 3; **8**, fruit × ¹/₂. *FERNANDOA ABBREVIATA* — **9**, leaflet × ¹/₂; **10**, detail of underside of leaflet × 10; **11**, inflorescence × ¹/₂; **12**, ovary × 3; **13**, anther × 3. *FERNANDOA LUTEA* — **14**, leaflet × ¹/₂; **15**, detail of underside of leaflet × 10; **16**, inflorescence × ¹/₂; **17**, ovary × 3; **18**, anther × 3. 1–3 from *Semsei* 1367, 4 from *Greenway* 9631, 5–7 from *Milne-Redhead & Taylor* 7332; 8 from *Gilbert & Kuchar* 5867, 9–13 from *Chapman & Tawakali* 6383, 14–15 from *Eggeling* 6044a, 16 from *Bryce in Eggeling* 6044 & *Schlieben* 782a (photo), 17–18 from *Eggeling* 6044. Drawn by Eleanor Catherine. Reproduced with permission from K.B. 49: 381–390 (1994).

adnate to the corolla for up to 4 mm; thecae 6–7.5 mm long, divaricate; staminode ± 3.5 mm long, linear, slightly expanded at the apex; ovary densely tomentellous, 9–11 mm long; style 4.5–6 cm long, stigma lobes elliptic, 3–4 × 3–4 mm, disk thick, cupular. Fruit not seen. Fig. 10.14–10.18, p. 41.

TANZANIA. Lindi District: Rondo Plateau, 25 Oct. 1934, *Schlieben* 5545! & Feb. 1951, from same tree as type, *Eggeling* 6044a! & SW end of Rondo Plateau, 13 km NE of Mihimi, 12 Nov. 1988, *Mackinder & Lock* 43!
DISTR. **T** 8; known only from Rondo Plateau
HAB. Semi-evergreen forest with *Milicia, Albizia, Dialium* and *Pteleopsis* on grey leached sandy soils; 600–750 m

SYN. *F. magnifica* Seem. var. *lutea* Verdc. in K.B. 7: 364 (1953)

2. **Fernandoa magnifica** *Seem.* in J.B. 8: 280 (1870) (as '*Ferdinandoa*') & in J.B. 9: 81 (1871); Sprague in F.T.A. 4(2): 517 (1906); T.T.C.L.: 70 (1949) pro parte; K.T.S.; 59, t. 2 (1961); Diniz in F.Z. 8(3): 81, t. 17 (1988) (as "*magnificia*") & in Fl. Moçamb. 120, Bignon.: 28 (1990); K.T.S.L.; 590, fig., map (1994); Bidgood in K.B. 49: 385, fig. 1/A–H (1994). Type: Tanzania, Masasi District: Rovuma R., Lake Chidia, *Kirk* s.n. (K!, holo.)

Shrub or small to large tree 2–20 m* tall; bark brown or grey, rough (*Smith* 1311 states smooth), longitudinally fissured, with prominent yellowish white lenticels. Leaves (4–)5–7-jugate, 18–30 cm long, 12–23 cm wide; rachis 5–25 cm long; petiole (1–)2.5–9 cm long; leaflets ovate to oblong or lanceolate, 3–14(–16) × 2–6 cm (to 19.5 × 8 cm in saplings), broadly cuneate to rounded at the base, entire or less often irregularly toothed to crenulate, attenuate to long acuminate at the apex, glabrous save for the domatia on lower surface; terminal petiolule up to 2.8 cm long but lateral leaflets sessile or almost so. Flowers in precocious racemes 6–10 cm long with thick peduncles (4–)5–17(–21) cm long; pedicels 5–7(–10) cm long, bent round like a walking stick below calyx. Calyx campanulate, 1.3–2.3 cm long, irregularly 3–5-lobed; lobes broadly ovate-triangular, with scattered small glands. Corolla orange-red or crimson with yellow throat which is ± crimson-lined, 5–11 cm long, 5-lobed, the lobes unequal, broadly rounded; filaments up to 6.5 cm long, adnate to corolla for up to 8 mm; thecae 9–12 mm long; staminode 2.5 mm long. Ovary cylindrical, 0.6–1.8 cm long, glabrous; style 4–8 cm long; stigma lobes elliptic, 2–3 mm long. Fruit terete, 33–54 cm long, 1–2 cm wide, spirally twisted and curved; seeds 7–12 mm long, 1.8–3.6 cm wide. Fig. 10.1–10.8, p. 41.

KENYA. Kwale District: near Jadini, 6 Dec. 1959, *Greenway* 9631!; Kilifi District: 9 km NE of Kaloleni, Chonyi Rock, 25 Nov. 1971, *Bally & A.R. Smith* 14392!; Lamu District: Boni Forest, near Kiunga, Oct. 1956, *Rawlins* 165!
TANZANIA. Handeni District: about 10 km on Korogwe–Handeni road, 19 Nov. 1955, *Milne-Redhead & Taylor* 7332!; Uzaramo District: Kurekese Forest Reserve, Kisiju, Sept. 1953, *Semsei* 1367!; Lindi District: Mahiwa Farm, 22 Sept. 1954, *F.G. Smith* 1311!
DISTR. **K** 7; **T** 3, 6, 8; Malawi, Mozambique & Mozambique-Zimbabwe border
HAB. Coastal thicket, woodland and *Afzelia* etc. forest; 5–450 m

SYN. *Heterophragma longipes* Bak. in K.B. 1894: 31 (1894). Types: Kenya, Mombasa & Ribe, *Wakefield* s.n. (K!, syn.) & Tanzania, Masasi District: Rovuma R., Lake Chidia, *Kirk* s.n. (K!, syn.)
Ferdinandia magnifica (Seem.) Sprague in F.T.A. 4(2): 517 (1906); Milne-Redh. in K.B. 3: 171 (1948)
Kigelia lanceolata sensu Dale, T.S.K.: 160 (1936), *non* Sprague

NOTE. This has been cultivated as an ornamental in Tanzania (Lushoto District: Amani Lab. Plantation 17, 9 Dec. 1932, *Greenway* 3299)

* Diniz's reference to 30 m probably refers to *F. lutea*

3. **Fernandoa abbreviata** *Bidgood* in K.B. 49: 386 (1994). Type: Malawi, Limbe, cultivated at entrance to Imperial Tobacco Group (grown from seed collected at Sambani Forest Reserve by Topham), *Chapman & Tawakali* 6383 (K!, holo; BR, FHO, MAL, iso.)

Tree 8–18 m tall; bark grey; young branches brownish, striate with scattered small glands, older branches grey-brown, striate, lenticellate, leaves 3–5-jugate, 20–36 × (17–)20–25(–30) cm, including petiole of up to 10 cm, rachis striate, puberulous; leaflets at least twice as long as broad, oblong to elliptic, (6–)8.5–16.5 × (2–)3.5–6 cm including petiolule of up to 3 mm long, rounded and frequently unequal at the base, margins repand to serrate, obtuse to acuminate at the apex, the lowermost pair frequently cordate; terminal elliptic, (6–)9–12 × 4.5–4.7 cm; young leaflets glandular and with scattered hairs on both surfaces, mature leaflets with a few scattered large glands particularly toward the base, with hairs restricted to the midrib above and to the veins beneath. Flowers borne singly or in few-flowered inflorescences either from the axils of the leaves or cauliflorous, usually with the leaves (but see note below); peduncles where present 0.3–2.2 cm long, minutely hairy; bracteoles caducous, 3–4 × 1–2 mm; pedicels 2–5 cm long, minutely hairy. Calyx broadly campanulate, 2–3(–3.5) × 2.5–3.5(–4) cm, 2–3-lobed, minutely hairy, chartaceous; lobes elliptic to broadly elliptic, margins scarious. Corolla dark maroon, reddish brown, orange-red or scarlet, yellow in the throat, broadly campanulate, (5–)6–8(–8.5) × (5–)6–10(–11) cm; lobes broadly elliptic to obovate to cordiform; tube 3.5–4 cm long; filaments 3.5–4.5 cm long, adnate to the corolla for ± 5 mm in the constricted base of the tube; thecae 5–8 mm long, divaricate; staminode 0.5–1 cm long, cupular; ovary ± 1 cm long, usually pubescent with greyish white hairs (see note); style ± 5 cm long; stigma lobes 2–2.5 × 2–2.5 mm, rounded at the apex. Disk thick, cupular. Fruit up to 60 cm long, 1–1.2 cm wide, slightly striate, minutely glandular; seeds not seen. Fig. 10.9–10.13, p. 41.

TANZANIA. Mpwapwa District: E Mpwapwa, 27 Aug. 1930, *Greenway* 2474!; Morogoro District: possibly near Morogoro itself, 1937, *Rounce* 608!; Iringa District: N Gologolo Mts., 12 Sept. 1970, *Thulin & Mhoro* 919!
DISTR. **T** 5–7; Malawi
HAB. Riverine fringing *Acacia* forest, *Combretum schumanii–Strychnos* forest, deciduous thicket, montane forest; 1000–1350 m

SYN. *F. magnifica* sensu T.T.C.L: 70 (1949) pro parte, *non* Seem.

8. **KIGELIA**

DC. in Bibl. Univ. Genève, sér. 2, 17: 135 (1838) & in Ann. Sci. Nat. sér. 2 Bot. 11: 297 (1839)

Tree; young branches with gland fields often present at the nodes. Leaves opposite or ternate, imparipinnate; leaflets (2–)4–6(–7) pairs. Inflorescence a pendulous raceme with spreading branches; pedicels bent upwards. Calyx tubular, 2-lipped. Corolla large, broadly campanulate, tube cylindrical at base, 5-lobed; stamens 4, in two unequal pairs, thecae free for most of their length, staminode present. Disk large, annular; ovary terete, one-celled with 2 parietal placentas; ovules numerous, multiseriate; stigma bilobed. Fruit large, cylindrical (sausage-shaped), indehiscent with thick walls. Seeds numerous, inserted in a tough fibrous pulp, not winged.

A single very variable species across tropical Africa. Widely cultivated elsewhere in the tropics.

Kigelia africana (*Lam.*) *Benth.* in Niger Fl.: 463 (1849); Sprague in F.T.A. 4(2): 536 (1906); T.T.C.L.: 71 (1949); Heine in F.W.T.A. ed. 2, 2: 385, fig. 294 (1963); Haerdi in Acta Trop. Suppl. 8: 148 (1964); Merxm. & Schreiber in Prodr. Fl. SW. Afr. 128: 3 (1967); Palmer & Pitman, Trees S. Afr. 3: 2011, figs. & photos. (1973); Liben, F.A.C.

Bignon.: 4, t. 1 (1977); Vollesen in Opera Bot.: 59: 77 (1980); Hamilton, Uganda Forest Trees: 203 (1981); Palgrave, Trees S. Afr.: 833 (1981); Gentry in Fl. Cameroun 27: 32, t. 9 (1984) & in Fl. Gabon 27: 27, t. 5 (1985); Blundell, Wild Fl. E. Afr.: 381, t. 485 (1987); Diniz in F.Z. 8(3): 83, t. 18 (1988); K.T.S.L.: 591, fig., map (1994); Tardelli & Settesoldi in Fl. Som. 3: 303, fig. 209 (2006). Type: Senegal, *Adanson* 199A (P-JUSS 4991, holo.; microfiche!)*

Shrub 2–3 m tall or tree often with wide spreading crown, 2.5–18(–24) m tall and possibly larger (but reports of up to 35 m incorrect); bark usually grey, occasionally brown, smooth to rough or ridged, scaly or flaking. Leaves opposite or ternate; leaflets thin to very coriaceous, (2–)3–8-jugate, oblong, elliptic, ovate or obovate, (6–)10–20(–30) × (4–)6–13(–16) cm, acute to cuneate or rounded, truncate, or emarginate at the base, entire to serrate (particularly sucker shoots), rounded, obtuse or acute to distinctly acuminate at the apex, glabrous to pubescent or tomentellous, smooth or often slightly to very scabrid with raised pale dots, peltate and punctate glands sometimes present; venation plane or sometimes raised or impressed, with 7–12 pairs of veins. Flowers said to smell unpleasant, in lax terminal hanging panicles 30–80(–150) cm long including peduncle up to 40 cm long; bracts lanceolate, 1 cm long; individual cymes 1–3-flowered; pedicels hooked, 1–4(–5) cm long. Calyx campanulate, 2–5 cm long, ± 2-lipped, irregularly lobed, the lobes ± 1 cm long, tomentose to glabrous, with scattered glands. Corolla at first with tube yellow outside, turning orange or red and with dark red to blackish inside, 3–9 cm long, cylindrical at base for a very variable length; lobes yellow-green outside suffused with crimson, dark crimson inside, ± ovate, 3–4.5 cm long and wide. Fruits cylindrical, 30–90 cm long, 7.5–10 cm wide, tough and woody; stalks up to 50 cm long; seeds ovoid, ± 11 mm long, 7 mm wide and 4 mm thick. Fig. 11, p. 45.

subsp. **africana**

Leaflets ± thin to very coriaceous, glabrous to tomentellous, usually slightly to strongly scabrid, rounded at the apex to shortly abruptly acuminate from a rounded apex. Essentially a plant of grassland with scattered trees or woodland.

UGANDA. Acholi District: Gulu–Pakwach road, km 48, 10 Dec. 1962, *Styles* 259!; Karamoja District: Kakumongole, 10 Jan. 1937, *A.S. Thomas* 2242!; Mbale District: Bulucheke–Tororo road, km 24, 14 Jan. 1963, *Styles* 317!
KENYA. Northern Frontier District: Horr Valley, 1962, *Stevens* 3!; Kisumu-Londiani District: Songhor, 28 Dec. 1958, *Dale* K 1034!; Kilifi District: Kibarani, 5 Mar 1945, *Jeffery* 111!
TANZANIA. Mwanza District: W Serengeti, 123 km from Mwanza on Musoma road, 17 July 1960, *Verdcourt* 2897!; Tanga District: Tanga, 12 Mar. 1907, *Braun* in A.H. 8110!; Mpwapwa District: E Mpwapwa, 25 Aug. 1930, *Greenway* 2465!
DISTR. **U** 1–3; **K** 1–7; **T** 1–8; **Z**, **P**; from Senegal to Sudan, Somalia and Ethiopia south to Botswana and eastern South Africa; also cultivated in the tropics elsewhere
HAB. Grassland with scattered trees, woodland, lake shores and riverine forest/woodland edges; 0–1950 m

SYN. *Bignonia africana* Lam., Encyl. Méth. 1: 424 (1785)
　　　Crescentia pinnata Jacq., Collect. 3: 203, t. 18 (1789). Type: Material from Mozambique via Mauritius, cult. at Hort. Schönbrunn, *Jacquin* (W, holo.?)
　　　Sotor aethiopum Fenzl, Ber. 21ste Versamml. Deutsch Naturf. 1843: 168 (1844). Type: Sudan, Fazogl [Fassokel], *Kotschy* 403 (W, holo., K!, iso.)
　　　Kigelia pinnata (Jacq.) DC., Prodr. 9: 247 (1845): Sprague in F.T.A. 4(2): 537 (1906); U.O.P.Z: 319 (1949); T.T.C.L: 71 (1949); K.T.S.: 60 (1961); F.F.N.R.: 379 (1962)

* Several authors have suggested Thouin was the collector because Lamarck mentions him but Thouin was a gardener at the Jardin des Plantes in Paris and he never went to Africa; he was a friend of Adanson and possibly received material from him.

FIG. 11. *KIGELIA AFRICANA* — **1**, inflorescence × ¹/₂; **2**, leaf × ¹/₂; **3**, flower opened × ¹/₂; **4**, fruit × ¹/₂; **5**, seed × ¹/₂. 1 from *Friis* 2504; 2, 3 from *Schweinfurth* 963/855; 4 from *Friis* 7085; 5 from *Schweinfurth* 1346. Drawn by Emmanuel Papadopoulos.

K. aethiopica Decne. in Deless., Ic. Select. 5: 39, t. 93 (1846); P.O.A. C: 364 (1895); Sprague in F.T.A. 4(2): 538 (1906); U.O.P.Z.: 320 (1949); T.T.C.L.: 71 (1949); I.T.U. ed. 2: 38 (1952); Jex-Blake, Gard. E. Afr.: 222, 335 (1957). Type: 'Ethiopia' but gives native names in Fazogl and Cordofan both in Sudan, t. 93 (lecto.!)

K. abyssinica A.Rich., Tent. Fl. Abyss. 2: 60 (1850), t. 75 (1851). Type: Ethiopia, R. Mareb, *Quartin Dillon* (P, holo.)

K. pinnata (Jacq.) DC. var. *tomentella* Sprague in F.T.A. 4(2): 537 (1906). Type: Zimbabwe, Victoria Falls, *Allen* 30 (K!, lecto., LISC, SRGH, photo., chosen by Diniz)

K. aethiopica Decne. var. *abyssinica* (A.Rich.) Sprague in F.T.A. 4(2): 538 (1906)

K. aethiopica Decne. var. *bornuensis* Sprague in F.T.A. 4(2): 538 (1906). Type: Nigeria, Bornu, *Vogel* 83 (K, syn!)

K. aethiopica Decne. var. *usambarica* Sprague in F.T.A. 4(2): 538 (1906); T.T.C.L.: 71 (1949). Type: Tanzania, Usambara, between Bwiti [Buiti] and M'lalo, *Holst* 2404 (K!, holo.)

K. talbotii Hutch. & Dalz., F.W.T.A. 2: 238 (1931). Type: Nigeria, Oban, *Talbot* 1354 (K!, holo., MO, photo.) (intermediate)

K. aethiopum (Fenzl) Dandy in F.P.S. 3: 156, fig. 42 (1956). Note: the epithet has often been mis-spelled as *aethiopium*; it is however a noun in apposition

K. moosa sensu Vollesen in Opera Bot. 59: 77 (1980), *non* Sprague (intermediate)

NOTE. Many specimens from **T** 5 and **T** 7 have densely pubescent or tomentellous leaflets particularly when young and correspond exactly with *K. pinnata* var. *tomentella* Sprague; this form is common in the F.Z. area but scattered specimens occur throughout the range of the subspecies. Some specimens of subsp. *africana* have the upper leaf surface very roughly scabrid and this form has frequently been used as sandpaper and corresponds to *K. aethiopica* var. *usambarica* Sprague. Occasionally both characters are found in one individual, e.g. *Mapunda & Raya* DSM 1046 (Dodoma District: Bahi, 16 Dec. 1969). *Eggeling* 6717 (Tanzania, Lindi District, Lake Lutamba, Nov. 1953) has greenish yellow flowers.

subsp. **moosa** (*Sprague*) *Bidgood & Verdc.* **comb. nov**. Type: Uganda, Sese Is., *Dawe* 1 & 63 and W Ankole, *Dawe* 412 (all K!, syn.)

Leaflets thin, never coriaceous, glabrous or slightly pubescent, not or slightly scabrid (see note), narrowed gradually into a long acumen. Essentially a plant of evergreen or swamp forest.

UGANDA. Toro District: Semliki Valley, Bwamba Forest, Feb. 1930, *Brasnett* F.D. 6!; Mbale Distrct: Mt Elgon, 22 Mar. 1924, *Snowden* 866!; Masaka District: Sese Is., Bugala, 7 Feb. 1934, *A.S. Thomas* 1227!

KENYA. South Nyeri District: Aberdare Forest, Gura Valley, Mar. 1930, *Gardner* in F.D. 2298!; Meru District: Meru, Upper Imenti Forest, 28 June 1974, *R.B. & A.J. Faden* 74/907!; North Kavirondo District: Kakamega Forest, near Forest Office, 9 Dec. 1956, *Verdcourt* 1678!

TANZANIA. Bukoba District: Minziro Forest, Mar. 1953, *Procter* 156! & Kaigi, 1935, *Gillman* 405!; Mpanda District: Kungwe–Mahali Peninsula, below Kungwe Mt, head of Mtali R., 7 Sept. 1959, *Harley* 9535!

DISTR. **U** 2–4; **K** 4, 5; **T** 1, 4; W Africa, Congo-Kinshasa, Sudan, Angola

HAB. Evergreen forest, swamp forest, upland bushland, in spray of waterfalls; 1050–2250 m

SYN. *K. moosa* Sprague in F.T.A. 4 (2): 536 (1906); F.P.N.A. 2: 252, fig. 11 (1947); I.T.U. ed. 2: 38, fig. 7 (1952); Jex-Blake, Gard. E. Afr. ed. 4: 335 (1957); K.T.S.: 60, fig. 12 (1961); K.T.S.L.: 591, map (1994)

K. lanceolata Sprague in F.T.A. 4 (2): 534 (1906); F.P.N.A. 2: 251 (1947); I.T.U. ed. 2: 38 (1952). Type: Uganda, Ruwenzori, *Scott Elliot* 7905 (K!, holo., BM!, iso.)

? *K. acutifolia* Sprague in F.T.A. 4 (2): 535 (1906), Cameroon, Bipinde, *Zenker* 1316 (K!, lecto., BM!, P, iso., MO photo) (chosen by Gentry)

K. angolensis Sprague in F.T.A. 4 (2): 535 (1906). Type: Angola, Golungo Alto, by streams, Sobatos de Gumba Bango Aquitamba and Sarge, *Welwitsch* 489 and near Trombeta, by R. Muia, *Welwitsch* 491 (both K!, syn., BM!, iso.)

K. elliptica Sprague in F.T.A. 4(2): 534 (1906): Type: Nigeria, Bonny, *Monteiro* s.n. (K!, lecto., MO, photo.), chosen by Gentry

K. impressa Sprague in F.T.A. 4 (2): 535 (1906) pro parte. Type: Fernando Po, *Barter* s.n. (K!, lecto., MO, photo.) (chosen here).

K. elliotii Sprague in F.T.A. 4 (2): 536 (1906); F.P.N.A. 2: 252 (1947). Type: Sierra Leone, Talla, near Kundita, *Scott Elliott* 5037 (K!, syn.) & Scarcies, near Baya, *Scott Elliott* 4757 (K!, syn.)

K. spragueana Wernham in J.B. 52: 31 (1914)*. Type: Nigeria, Eket, near Mkpokk, *Talbot* 3392 (BM!, holo.)

NOTE. Since in East Africa two taxa have been kept up by workers who could by no means be termed splitters we have not followed Heine, Liben and Diniz in considering the species indivisible but recognised two 'ecological' subspecies. Those who prefer not to recognise two taxa can ignore the subspecies and those who wish to use them can do so. We agree there are too many intermediates to make recognition of two species appropriate. *Vollesen* MRC2906 (Tanzania, Selous Game Reserve, Ruaha Camp, 15 km E of Kilombero Sugar Estate) is such an intermediate with only shortly acuminate leaves. Subsp. *moosa* usually has shorter inflorescences and smaller flowers with corolla tube 5–8 cm long and lobes about 2 cm long and wide. Field workers have always emphasised differences in flower colour, those of subsp. *moosa* being more yellow and orange but trying to make generalisations from the numerous colour notes available has not supported this. Our own observations for subsp. *africana* are buds: dull crimson red with veins yellow-green; open flowers: base of corolla tube greenish cream, upper part of tube dull crimson-red outside with greenish yellow veins, outside of lobes mostly yellow-green but suffused with crimson; inside of corolla liver-coloured with some greenish cream marks. For subsp. *moosa* corolla at first yellow with lines of salmon spots so close as to form continuous lines, later turning orange but becoming crimson-red when dead. Material of subsp. *moosa* from Kakamega and Malaba grown at T.H. Jackson's garden at Kapretwa near Kitale on Mt Elgon (*Bally* 4828, *Tweedie* 3063, June 1965) retains the acuminate leaves and small flowers in cultivation.

Rendle 586 (Kenya, near E slope of Mt Kenya, 1800 m, 27 Aug. 1929) has acuminate very scabrid leaves and slender flowers; we have treated it as a form of subsp. *moosa* but it is very distinctive.

---

* Hutchinson & Dalziel (FWTA 2: 240 (1931)) sink this into *K. elliptica* Sprague, giving 'tree 20–30 ft' but Wernham gives 80 ft. The Talbots were unlikely to be wrong and this must be one of the tallest specimens recorded for the species as a whole.

Fig. 12. *COBAEA SCANDENS* — **1**, vegetative shoot × ¹/₂; **2**, flower × ¹/₂; **3**, fruit × ¹/₂; **4**, seeds × ¹/₂. 1, 2 from *Polhill & Paulo* 1753; 3, 4 from *Grimshaw* 93/111. Drawn by Emmanuel Papadopoulos.

FLORA OF TROPICAL EAST AFRICA

COBAEACEAE

BERNARD VERDCOURT

Herbaceous or shrubby perennial climbers. Leaves alternate, pinnate, stipulate, the stipules large and foliaceous, the terminal leaflet modified into a divided tendril. Inflorescences axillary, 1–3-flowered, long pedunculate, pendent. Calyx foliaceous, 5-lobed, reflexed. Corolla campanulate, 5-lobed, the lobes rounded to linear. Stamens 5, exserted, epipetalous; filaments bearded beneath; anthers versatile. Ovary 3-locular with 2-numerous axile ovules per locule; style long, shortly trifid at apex; disk large, fleshy. Capsule 3-valved, septicidal. Seeds biseriate, medifixed, compressed, winged.

One genus native in the New World.

COBAEA

Cav., Icones Descr. Pl. Hisp. 1: 11, t. 16, 17 (1791)

Description as for family.

About 20 species, Mexico to tropical South America. Usually referred to the Polemoniaceae, but it bears little resemblance to that family. Airy Shaw (Willis, Dict. Fl. Pl. ed. 8: 265 (1973), kept the family Cobaeaceae separate. The Angiosperm Phylogeny Group include *Cobaea* in the subfamily Cobaeoideae of Polemoniaceae together with several other disparate looking genera.

Cobaea scandens *Cav.*, Icones Descr. Pl. Hisp. 1: 11, t. 16, 17 (1791); Jex-Blake, Gard. E. Afr. ed. 4: 134 (1957); Ruffo et al., Cat. Lushoto Herb.: 273 (1996). Type: specimen grown in the Royal Garden of Madrid from material collected near Mexico City (MA-475555, holo.; microfiche!)

Climber 2–9 m tall. Leaflets 4–6 per leaf, oblong or elliptic, 3–10 cm long, 1.5–5 cm wide, acute; stipules 3–10.5 cm long, 1.5–4 cm wide, panduriform; tendrils hook-shaped. Peduncles up to 2.5 cm long, the pedicel twisted after flowering. Flowers erect in bud but becoming horizontal or pendent. Calyx-lobes ± round, 2.5–3 cm long, 2–2.5 cm wide. Corolla greenish cream and musk-scented at first, later violet to deep purple and sweet-scented, 5–6 cm long, 4–5 cm wide; disk crinkly. Capsule valves thinly woody, elliptic, 5–7 cm long, 2–3 cm wide. Fig. 12, p. 48.

TANZANIA. Moshi District: Kilimanjaro, Kilimanjaro Timbers, 3 June 1993, *Grimshaw* 93/111!; Lushoto District: Lushoto Camphor Forest, 9 June 1977, *Mtui et al.* 8!; Iringa District: Mufindi, Kigogo Forest Reserve, Aug. 1953, *Carmichael* 250!
DISTR. **T** 2, 3, 7; native of Mexico (Puebla) but now naturalised in Tanzania and cultivated in Kenya highlands and widely elsewhere, often becoming naturalised
HAB. Evergreen forest at edges and in disturbed areas; 1500–2000 m

NOTE. The earliest Tanzanian specimen of this seen is *Hughes* 43, Lushoto District: Bushbuck Valley, collected 14 May 1947.
Note on the genus *Phlox* L. (Polemoniaceae): species of this are grown in East Africa. Jex-Blake

(Gard. E. Afr., ed. 4: 67, 80, 360 (1957)) mentions *P. drummondii* Hook. a widely grown annual garden plant with many varieties and *P. decussata* Pursh (correctly *P. paniculata* L. a perennial). Ruffo et al. (Cat. Lushoto Herb.: 273 (1996) mentions a *Phlox* sp. *Shabani* 930 from Lushoto. **Adenocalymma marginatum** (*Cham.*) *DC.*, 8

INDEX TO BIGNONIACEAE

New names validated in this part

Kigelia africana (*Lam.*) *Benth.* subsp. **moosa** (*Sprague*) *Bidgood & Verdc.* comb. nov.
Tecomaria capensis (*Thunb.*) *Spach* var. **flava** Verdc. var. nov.

INDEX TO COBAEACEAE

No new names validated in this part

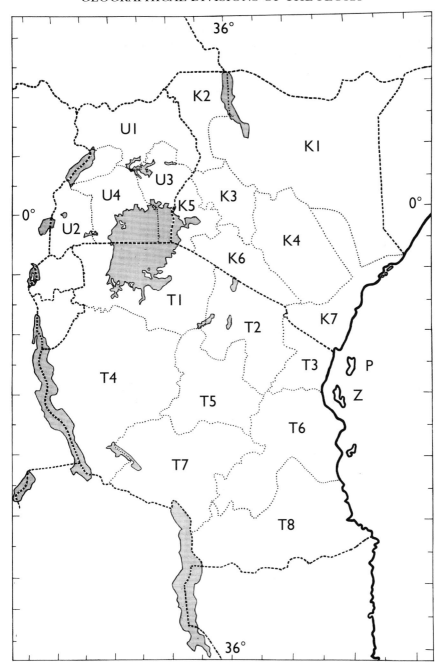

PLANTS PEOPLE
POSSIBILITIES

First published in 2006 by
Royal Botanic Gardens, Kew
Richmond, Surrey, TW9 3AB, UK
www.kew.org

ISBN 1 84246 151 6

British Library Cataloguing in Publication Data
A catalogue record for this book is available from the British Library

Design and typesetting by Margaret Newman,
Kew Publishing, Royal Botanic Gardens, Kew.

Printed in the UK by Hobbs the Printers

For information or to purchase all Kew titles please visit
www.kewbooks.com or email publishing@kew.org

All proceeds go to support Kew's work in saving the world's plants for life